Criteria for a Recommended Standard

Occupational Exposure to Hexavalent Chromium

DEPARTMENT OF HEALTH AND HUMAN SERVICES
Centers for Disease Control and Prevention
National Institute for Occupational Safety and Health

On the cover left to right, top to bottom: (1) Spray painting aircraft with chromate paint, (2) Electroplating tank, (3) Welder, and (4) Working with Portland cement. Images courtesy of U.S. Navy, NIOSH Division of Applied Research and Technology, and Thinkstock. Used with permission.

> This document is in the public domain and may be freely copied or reprinted

Disclaimer

Mention of any company or product does not constitute endorsement by the National Institute for Occupational Safety and Health (NIOSH). In addition, citations to Web sites external to NIOSH do not constitute NIOSH endorsement of the sponsoring organizations or their programs or products. Furthermore, NIOSH is not responsible for the content of these Web sites.

Ordering Information

This document is in the public domain and may be freely copied or reprinted. To receive NIOSH documents or other information about occupational safety and health topics, contact NIOSH at

   Telephone: 1–800–CDC–INFO (1–800–232–4636)
   TTY: 1–888–232–6348
   E-mail: cdcinfo@cdc.gov

or visit the NIOSH Web site at www.cdc.gov/niosh.

DHHS (NIOSH) Publication No. 2013–128
(Revised with minor technical changes)

September 2013

SAFER • HEALTHIER • PEOPLE™

Foreword

When the U.S. Congress passed the Occupational Safety and Health Act of 1970 (Public Law 91-596), it established the National Institute for Occupational Safety and Health (NIOSH). Through the Act, Congress charged NIOSH with recommending occupational safety and health standards and describing exposure levels that are safe for various periods of employment, including but not limited to the exposures at which no worker will suffer diminished health, functional capacity, or life expectancy because of his or her work experience.

Criteria documents contain a critical review of the scientific and technical information about the prevalence of hazards, the existence of safety and health risks, and the adequacy of control methods. By means of criteria documents, NIOSH communicates these recommended standards to regulatory agencies, including the Occupational Safety and Health Administration (OSHA), health professionals in academic institutions, industry, organized labor, public interest groups, and others in the occupational safety and health community.

This criteria document is derived from the NIOSH evaluation of critical health effects studies of occupational exposure to hexavalent chromium (Cr[VI]) compounds. It provides recommendations for controlling workplace exposures including a revised recommended exposure limit (REL) derived using current quantitative risk assessment methodology on human health effects data. This document supersedes the 1975 *Criteria for a Recommended Standard: Occupational Exposure to Chromium(VI)* and *NIOSH Testimony to OSHA on the Proposed Rule on Occupational Exposure to Hexavalent Chromium* [NIOSH 1975a, 2005a].

Cr(VI) compounds include a large group of chemicals with varying chemical properties, uses, and workplace exposures. Their properties include corrosion-resistance, durability, and hardness. Sodium dichromate is the most common chromium chemical from which other Cr(VI) compounds may be produced. Materials containing Cr(VI) include various paint and primer pigments, graphic art supplies, fungicides, corrosion inhibitors, and wood preservatives. Some of the industries in which the largest numbers of workers are exposed to high concentrations of Cr(VI) compounds include electroplating, welding, and painting. An estimated 558,000 U.S. workers are exposed to airborne Cr(VI) compounds in the workplace.

Cr(VI) is a well-established occupational carcinogen associated with lung cancer and nasal and sinus cancer. NIOSH considers all Cr(VI) compounds to be occupational carcinogens. NIOSH recommends that airborne exposure to all Cr(VI) compounds be limited to a concentration of 0.2 µg Cr(VI)/m^3 for an 8-hr time-weighted average (TWA) exposure, during a 40-hr workweek. The REL is intended to reduce workers' risk of lung cancer associated with occupational exposure to Cr(VI) compounds over a 45-year working lifetime. It is expected that reducing airborne workplace exposures to Cr(VI) will also reduce the nonmalignant respiratory effects of Cr(VI) compounds, including irritated, ulcerated, or perforated nasal septa and other potential adverse health effects. Because of the residual risk of lung cancer at the REL, NIOSH further recommends that continued efforts be made to reduce Cr(VI) exposures to below the REL. A hierarchy of controls, including elimination, substitution, engineering controls, administrative controls, and the use of personal protective equipment, should be followed to control workplace exposures.

In addition to limiting airborne concentrations of Cr(VI) compounds, NIOSH recommends that dermal exposure to Cr(VI) be prevented in the workplace to reduce the risk of adverse dermal effects, including irritation, corrosion, ulcers, skin sensitization, and allergic contact dermatitis. An estimated 1,045,500 U.S. workers have dermal exposure to Cr(VI) in cement, primarily in the construction industry.

NIOSH urges employers to disseminate this information to workers and customers. NIOSH also requests that professional and trade associations and labor organizations inform their members about the hazards of occupational exposure to Cr(VI) compounds.

NIOSH appreciates the time and effort taken by the expert peer, stakeholder, and public reviewers to provide comments on this document. Their input strengthened this document.

John Howard, MD
Director, National Institute for
  Occupational Safety and Health
Centers for Disease Control and Prevention

Executive Summary

In this criteria document, the National Institute for Occupational Safety and Health (NIOSH) reviews the critical health effects studies of hexavalent chromium (Cr[VI]) compounds in order to update its assessment of the potential health effects of occupational exposure to Cr(VI) compounds and its recommendations to prevent and control these workplace exposures. NIOSH reviews the following aspects of workplace exposure to Cr(VI) compounds: the potential for exposures (Chapter 2), analytical methods and considerations (Chapter 3), human health effects (Chapter 4), experimental studies (Chapter 5), and quantitative risk assessments (Chapter 6). Based on evaluation of this information, NIOSH provides recommendations for a revised recommended exposure limit (REL) for Cr(VI) compounds (Chapter 7) and other recommendations for risk management (Chapter 8).

This criteria document supersedes previous NIOSH Cr(VI) policy statements, including the 1975 *NIOSH Criteria for a Recommended Standard: Occupational Exposure to Chromium(VI)* and *NIOSH Testimony to OSHA on the Proposed Rule on Occupational Exposure to Hexavalent Chromium* [NIOSH 1975a, 2005a]. Key information in this document, including the NIOSH site visits and the NIOSH quantitative risk assessment, were previously submitted to the Occupational Safety and Health Administration (OSHA) and were publicly available during the OSHA Cr(VI) rule-making process. OSHA published its final standard for Cr(VI) compounds in 2006 [71 Fed. Reg. 10099 (2006)].

Cr(VI) compounds include a large group of chemicals with varying chemical properties, uses, and workplace exposures. Their properties include corrosion-resistance, durability, and hardness. Workers may be exposed to airborne Cr(VI) when these compounds are manufactured from other forms of Cr (e.g., the production of chromates from chromite ore); when products containing Cr(VI) are used to manufacture other products (e.g., chromate-containing paints, electroplating); or when products containing other forms of Cr are used in processes that result in the formation of Cr(VI) as a by-product (e.g., welding). In the marketplace, the most prevalent materials that contain chromium are chromite ore, chromium chemicals, ferroalloys, and metal. Sodium dichromate is the most common chromium chemical from which other Cr(VI) compounds may be produced. Cr(VI) compounds commonly manufactured include sodium dichromate, sodium chromate, potassium dichromate, potassium chromate, ammonium dichromate, and Cr(VI) oxide. Other manufactured materials containing Cr(VI) include various paint and primer pigments, graphic arts supplies, fungicides, and corrosion inhibitors.

An estimated 558,000 U.S. workers are exposed to airborne Cr(VI) compounds in the workplace. Some of the industries in which the largest numbers of workers are exposed to high concentrations of airborne Cr(VI) compounds include electroplating, welding, and painting. An estimated 1,045,500 U.S. workers have dermal exposure to Cr(VI) in cement, primarily in the construction industry.

Cr(VI) is a well-established occupational carcinogen associated with lung cancer and nasal and sinus cancer. NIOSH considers all Cr(VI) compounds to be occupational carcinogens [NIOSH 1988b, 2002, 2005a]. In 1989, the International Agency for Research on Cancer (IARC) critically evaluated the published epidemiologic studies of chromium compounds. IARC concluded that "there is sufficient evidence in humans for the carcinogenicity of chromium[VI] compounds as encountered in the chromate production, chromate pigment production and chromium plating industries" (i.e., IARC category "Group 1" carcinogen) [IARC 1990]. Cr(VI) compounds were reaffirmed as an IARC Group 1 carcinogen (lung) in 2009 [Straif et al. 2009; IARC 2012]. The National Toxicology Program (NTP) identified Cr(VI) compounds as carcinogens in its first annual report on carcinogens in 1980 [NTP 2011]. Nonmalignant respiratory effects of Cr(VI) compounds include irritated, ulcerated, or perforated nasal septa. Other adverse health effects, including reproductive and developmental effects, have been reviewed by other government agencies [71 Fed. Reg. 10099 (2006); ATSDR 2012; EPA 1998; Health Council of the Netherlands 2001; OEHHA 2009].

Studies of the Baltimore and Painesville cohorts of chromate production workers [Gibb et al. 2000b; Luippold et al. 2003] provide the best information for predicting Cr(VI) cancer risks because of the quality of the exposure estimation, large amount of worker data available for analysis, extent of exposure, and years of follow-up [NIOSH 2005a]. NIOSH selected the Baltimore cohort [Gibb et al. 2000b] for analysis because it has a greater number of lung cancer deaths, better smoking histories, and a more comprehensive retrospective exposure archive. The NIOSH risk assessment estimates an excess lifetime risk of lung cancer death of 6 per 1,000 workers at 1 µg Cr(VI)/m^3 (the previous REL) and approximately 1 per 1,000 workers at 0.2 µg Cr(VI)/m^3 (the revised REL) [Park et al. 2004]. The basis for the previous REL for carcinogenic Cr(VI) compounds was the quantitative limitation of the analytical method available in 1975.

Based on the results of the NIOSH quantitative risk assessment [Park et al. 2004], NIOSH recommends that airborne exposure to all Cr(VI) compounds be limited to a concentration of 0.2 µg Cr(VI)/m^3 for an 8-hr TWA exposure, during a 40-hr workweek. The REL is intended to reduce workers' risk of lung cancer associated with occupational exposure to Cr(VI) compounds over a 45-year working lifetime. It is expected that reducing airborne workplace exposures to Cr(VI) will also reduce the nonmalignant respiratory effects of Cr(VI) compounds, including irritated, ulcerated, or perforated nasal septa and other potential adverse health effects. Because of the residual risk of lung cancer at the REL, NIOSH recommends that continued efforts be made to reduce exposures to Cr(VI) compounds below the REL.

The available scientific evidence supports the inclusion of all Cr(VI) compounds into this recommendation. Cr(VI) compounds studied have demonstrated their carcinogenic potential in animal, in vitro, or human studies [NIOSH 1988b; 2002; 2005a,b]. Molecular toxicology studies provide additional support for classifying Cr(VI) compounds as occupational carcinogens.

The NIOSH REL is a health-based recommendation derived from the results of the NIOSH quantitative risk assessment conducted on human health effects data. Additional considerations include analytical feasibility and the achievability of engineering controls. NIOSH Method 7605, OSHA Method ID-215, and international consensus standard analytical methods can quantitatively assess worker exposure to Cr(VI) at the REL. Based on a qualitative assessment of workplace exposure data, NIOSH acknowledges that Cr(VI) exposures below the REL can be achieved in some workplaces using existing technologies but are more difficult to control in others [Blade et al. 2007]. Some operations, including hard chromium electroplating, chromate-paint spray application, atomized-alloy spray-coating, and welding may have difficulty in consistently achieving exposures at or below the REL by means of engineering controls and work practices [Blade et al. 2007]. The extensive analysis of workplace exposures conducted for the OSHA rule-making process supports the NIOSH assessment that the REL is achievable in some workplaces but difficult to achieve in others [71 Fed. Reg. 10099 (2006)].

A hierarchy of controls, including elimination, substitution, engineering controls, administrative controls, and the use of personal protective equipment, should be followed to control workplace exposures. The REL is intended to promote the proper use of existing control technologies and to encourage the research and development of new control technologies where needed, in order to control workplace Cr(VI) exposures.

At this time, there are insufficient data to conduct a quantitative risk assessment for workers exposed to Cr(VI), other than chromate production workers or specific Cr(VI) compounds other than sodium dichromate. However, epidemiologic studies demonstrate that the health effects of airborne exposure to Cr(VI) are similar across workplaces and industries (see Chapter 4). Therefore, the results of the NIOSH quantitative risk assessment conducted on chromate production workers [Park et al. 2004] are used as the basis of the REL for all workplace exposures to Cr(VI) compounds.

The primary focus of this document is preventing workplace airborne exposure to Cr(VI) compounds to reduce workers' risk of lung cancer. However, NIOSH also recommends that dermal exposure to Cr(VI) compounds be prevented in the workplace to reduce adverse dermal effects including skin irritation, skin ulcers, skin sensitization, and allergic contact dermatitis.

NIOSH recommends that employers implement measures to protect the health of workers exposed to Cr(VI) compounds under a comprehensive safety and health program, including hazard communication, respiratory protection programs, smoking cessation, and medical monitoring. These elements, in combination with efforts to maintain airborne Cr(VI) concentrations below the REL and prevent dermal contact with Cr(VI) compounds, will further protect the health of workers.

Contents

Foreword iii
Executive Summary v
Abbreviations xiii
Acknowledgments xvii

1 Introduction 1
   1.1 Purpose and Scope 1
   1.2 History of NIOSH Cr(VI) Policy 1
   1.3 The REL for Cr(VI) Compounds 2

2 Properties, Production, and Potential for Exposure 5
   2.1 Physical and Chemical Properties 5
   2.2 Production and Use in the United States 5
   2.3 Potential Sources of Occupational Exposure 7
      2.3.1 Airborne Exposure 7
      2.3.2 Dermal Exposure 8
   2.4 Industries with Potential Exposure 8
      2.4.1 Airborne Exposure 8
      2.4.2 Dermal Exposure 8
   2.5 Number of U.S. Workers Potentially Exposed 8
   2.6 Measured Exposure in the Workplace 9
      2.6.1 NIOSH Multi-Industry Field Study [Blade et al. 2007] 9
      2.6.2 Shaw Environmental Report [2006] 11
      2.6.3 Welding and Thermal Cutting of Metals 21
   2.7 Occupational Exposure Limits 22
   2.8 IDLH Value 22
   2.9 Future Trends 22

3 Measurement of Exposure 25
   3.1 Air-Sampling Methods 25
      3.1.1 Air Sample Collection 25
      3.1.2 Air-Sampling Considerations 26
   3.2 Analytical Methods 26
      3.2.1 Cr(VI) Detection in Workplace Air 26
      3.2.2 Wipe Sampling Methods 27
   3.3 Biological Markers 29
      3.3.1 Biological Markers of Exposure 29

		3.3.2 Biological Markers of Effect	31

4 Human Health Effects . . . 33

4.1 Cancer . . . 33
- 4.1.1 Lung Cancer . . . 33
- 4.1.2 Nasal and Sinus Cancer . . . 38
- 4.1.3 Nonrespiratory Cancers . . . 39
- 4.1.4 Cancer Meta-Analyses . . . 39
- 4.1.5 Summary of Cancer and Cr(VI) Exposure . . . 40

4.2 Nonmalignant Effects . . . 41
- 4.2.1 Respiratory Effects . . . 41
- 4.2.2 Dermatologic Effects . . . 46
- 4.2.3 Reproductive Effects . . . 46
- 4.2.4 Other Health Effects . . . 47

5 Experimental Studies . . . 59

5.1 Pharmacokinetics . . . 59
5.2 Mechanisms of Toxicity . . . 59
5.3 Health Effects in Animals . . . 62
- 5.3.1 Subchronic Inhalation Studies . . . 62
- 5.3.2 Chronic Inhalation Studies . . . 64
- 5.3.3 Intratracheal Studies . . . 65
- 5.3.4 Intrabronchial Studies . . . 65
- 5.3.5 Chronic Oral Studies . . . 65
- 5.3.6 Reproductive Studies . . . 67

5.4 Dermal Studies . . . 67
- 5.4.1 Human Dermal Studies . . . 67
- 5.4.2 Animal Dermal Studies . . . 68
- 5.4.3 In Vitro Dermal Studies . . . 68

5.5 Summary of Animal Studies . . . 69

6 Quantitative Assessment of Risk . . . 71

6.1 Overview . . . 71
6.2 Baltimore Chromate Production Risk Assessments . . . 72
6.3 Painesville Chromate Production Risk Assessments . . . 75
6.4 Other Cancer Risk Assessments . . . 76
6.5 Summary . . . 77

7 Recommendations for an Exposure Limit . . . 79

7.1 The NIOSH REL for Cr(VI) Compounds . . . 79
7.2 Basis for NIOSH Standards . . . 79
7.3 Evidence for the Carcinogenicity of Cr(VI) Compounds . . . 81
- 7.3.1 Epidemiologic Lung Cancer Studies . . . 81

		7.3.2 Lung Cancer Meta-Analyses	82
		7.3.3 Animal Experimental Studies	83
	7.4	Basis for the NIOSH REL	83
		7.4.1 Park et al. [2004] Risk Assessment	85
		7.4.2 Crump et al. [2003] Risk Assessment	85
		7.4.3 Risk Assessment Summary	86
	7.5	Applicability of the REL to All Cr(VI) Compounds	86
	7.6	Analytical Feasibility of the REL	86
	7.7	Controlling Workplace Exposures	86
	7.8	Preventing Dermal Exposure	87
	7.9	Summary	87

8 Risk Management 89

- 8.1 NIOSH Recommended Exposure Limit 89
  - 8.1.1 The NIOSH REL 89
  - 8.1.2 Sampling and Analytical Methods 90
- 8.2 Informing Workers about the Hazard 90
  - 8.2.1 Safety and Health Programs 90
  - 8.2.2 Labeling and Posting 90
- 8.3 Exposure Control Measures 91
  - 8.3.1 Elimination and Substitution 91
  - 8.3.2 Engineering Controls 92
  - 8.3.3 Administrative Controls and Work Practices 97
  - 8.3.4 Protective Clothing and Equipment 97
- 8.4 Emergency Procedures 100
- 8.5 Exposure Monitoring Program 100
- 8.6 Medical Monitoring 102
  - 8.6.1 Worker Participation 102
  - 8.6.2 Medical Monitoring Program Director 103
  - 8.6.3 Medical Monitoring Program Elements 103
  - 8.6.4 Medical Reporting 105
  - 8.6.5 Employer Actions 105
- 8.7 Smoking Cessation 106
- 8.8 Record Keeping 106

References 109

Appendix A: Hexavalent Chromium and Lung Cancer in the Chromate Industry: A Quantitative Risk Assessment 127

Abbreviations

ACD	allergic contact dermatitis
ACGIH	American Conference of Governmental Industrial Hygienists
AIHA	American Industrial Hygiene Association
ACS	American Cancer Society
AlM	alveolar macrophage
AL	action level
APF	assigned protection factor
ASTM	American Society for Testing and Materials
ATSDR	Agency for Toxic Substances and Disease Registry
BAL	bronchoalveolar lavage
BEI	Biological Exposure Index
CCA	chromated copper arsenate
CI	confidence interval
CPC	chemical protective clothing
Cr	chromium
Cr(0)	metallic or elemental chromium
Cr(III)	trivalent chromium
Cr(VI)	hexavalent chromium
CrO_3	chromic acid or chromium trioxide
d	day
DECOS	Dutch Expert Committee on Occupational Standards
DNA	deoxyribonucleic acid
DPC	diphenylcarbazide/diphenylcarbazone
EID	Education and Information Division of the National Institute for Occupational Safety and Health
EPA	U.S. Environmental Protection Agency
EPRI	Electric Power Research Institute
FCAW	flux cored arc welding
Fed. Reg.	Federal Register
FEV_1	forced expiratory volume in one second

FEV$_1$/FVC	ratio of forced expiratory volume in one second (FEV$_1$) to forced vital capacity (FVC)
FVC	forced vital capacity
G2/M	gap 2/mitosis
GHS	Globally Harmonized System of Classification and Labelling of Chemicals
GM	geometric mean
GMAW	gas-metal arc welding
GTAW	gas-tungsten arc welding
GSD	geometric standard deviation
HCS	(OSHA) Hazard Communication Standard
HEPA	high-efficiency particulate air
hr	hour
H$_2$O$_2$	hydrogen peroxide
HIF-1	hypoxia-induced factor 1
HHE	Health Hazard Evaluation
IARC	International Agency for Research on Cancer
ICDA	International Chromium Development Association
IDLH	Immediately Dangerous to Life and Health
Ig	immunoglobulin
ILO	International Labour Organization
IMIS	Integrated Management Information System
ISO	International Organization for Standardization
IU	International Unit
l	liter
LDH	lactate dehydrogenase
LD$_{50}$	lethal dose resulting in 50% mortality
LEV	local exhaust ventilation
LH	luteinizing hormone
LOD	limit of detection
lnL	log-likelihood
M	molar
MCE	mixed cellulose ester
mg/m^3	milligrams per cubic meter of air
MIG	metal inert gas (welding)

mM	millimolar
MMA	manual metal arc (welding)
MMD	mass median diameter
MMAD	mass median aerodynamic diameter
MRL	minimum risk level
MSDS	Material Safety Data Sheet
n	number (sample size)
NAG	N-acetyl-β-D-glucosaminidase
nd	not detectable
ng	nanogram
nmol	nanomoles
Ni	nickel
NADPH	nicotinamide adenine dinucleotide phosphate
NIOSH	National Institute for Occupational Safety and Health
NOAEL	no observed adverse effect level
NOES	National Occupational Exposure Survey
NTP	National Toxicology Program
OEL	occupational exposure limit
·OH	hydroxyl radical
OR	odds ratio
OSHA	Occupational Safety and Health Administration
P	probability
PAPR	powered air-purifying respirator
PBZ	personal breathing zone
PCMR	proportionate cancer mortality ratio
PEL	Permissible Exposure Limit
PPE	personal protective equipment
ppm	parts per million
PVC	polyvinyl chloride
redox	reduction-oxidation
REL	Recommended Exposure Limit
ROS	reactive oxygen species
SD	standard deviation

SDS	Safety Data Sheet
SIC	Standard Industrial Classification
SMAW	shielded-metal arc welding
SMR	standardized mortality ratio
SOD	superoxide dismutase
T	tons
TIG	tungsten inert gas (welding)
TLV	Threshold Limit Value
TWA	time-weighted average
µg	microgram(s)
µg/l	microgram(s) per liter
µg/m^3	microgram(s) per cubic meter of air
µM	micromolar
U.K.	United Kingdom
U.S.	United States
UV-Vis	ultraviolet-visible
VEGF	vascular endothelial growth factor
WHO	World Health Organization
wk	week
yr	year(s)

Acknowledgments

This document was prepared by the Education and Information Division (EID), Paul Schulte, Director; Document Development Branch, T.J. Lentz, Chief; Risk Evaluation Branch, Christine Sofge, Chief. Kathleen MacMahon was the Project Officer. Faye Rice; Robert Park; Henryka Nagy (NIOSH retired); Leo Michael Blade (NIOSH retired); Kevin Ashley (NIOSH/DART); G. Kent Hatfield (NIOSH retired); and Thurman Wenzl (NIOSH retired) were major contributors.

For contributions to the technical content and review of this document, the authors gratefully acknowledge the following NIOSH personnel:

Division of Applied Research and Technology
James Bennett
Michael Gressel

Division of Respiratory Disease Studies
Lee Petsonk

Division of Surveillance, Hazard Evaluations, and Field Studies
Steven Ahrenholz
Douglas Trout

Education and Information Division
David Dankovic
Charles Geraci
Leslie Stayner
David Votaw
Ralph Zumwalde

Health Effects Laboratory Division
Vincent Castranova
Xianglin Shi
Stephen Leonard

Office of the Director
John Decker
Matthew Gillen
Paul Middendorf
Anita Schill

Pittsburgh Research Laboratory
Heinz Ahlers
Roland Berry Ann
Pengfei Gao
Bill Hoffman
Bob Stein
Doris Walter

The authors wish to thank Vanessa Williams for the document design and layout. Editorial and document assistance was provided by John Lechliter, Norma Helton, and Daniel Echt.

Special appreciation is expressed to the following individuals for serving as independent, external peer reviewers and providing comments that contributed to the development of this document:

Harvey J. Clewell, Ph.D., D.A.B.T.
Director, Center for Human Health Assessment
The Hamner Institutes for Health Sciences
Research Triangle Park, NC

Richard Danchik, Ph.D.
President, PittCon
Pittsburgh, PA

Herman J. Gibb, Ph.D., MPH
President, Tetra Tech Sciences
Arlington, VA

Edwin van Wijngaarden, Ph.D.
Associate Professor of Community and Preventive Medicine,
    Environmental Medicine, and Dentistry
Chief, Division of Epidemiology
University of Rochester School of Medicine and Dentistry

John Wise, Ph.D.
Professor, Toxicology and Molecular Epidemiology
Department of Applied Medical Sciences
Director, Maine Center for Toxicology and Environmental Health
University of Southern Maine

1 Introduction

1.1 Purpose and Scope

This criteria document describes the most recent NIOSH scientific evaluation of occupational exposure to hexavalent chromium (Cr[VI]) compounds, including the justification for a revised recommended exposure limit (REL) derived using current quantitative risk assessment methodology on human health effects data. This criteria document focuses on the relevant critical literature published since the 1975 *Criteria for a Recommended Standard: Occupational Exposure to Chromium(VI)* [NIOSH 1975a]. The policies and recommendations in this document provide updates to the *NIOSH Testimony on the OSHA Proposed Rule on Occupational Exposure to Hexavalent Chromium* and the corresponding *NIOSH Post-Hearing Comments* [NIOSH 2005a,b]. This final document incorporates the NIOSH response to peer, stakeholder, and public review comments received during the external review process.

1.2 History of NIOSH Cr(VI) Policy

In the 1973 *Criteria for a Recommended Standard: Occupational Exposure to Chromic Acid*, NIOSH recommended that the federal standard for chromic acid, 0.1 mg chromium trioxide/m^3, as a 15-minute ceiling concentration, be retained because of reports of nasal ulceration occurring at concentrations only slightly above this concentration [NIOSH 1973a]. In addition, NIOSH recommended 0.05 mg chromium trioxide/m^3 time-weighted average (TWA) for an 8-hour workday, 40-hour work week, to protect against possible chronic effects, including lung cancer and liver damage.

In the 1975 *Criteria for a Recommended Standard: Occupational Exposure to Chromium(VI)*, NIOSH supported two distinct recommended standards for Cr(VI) compounds [NIOSH 1975a]. Some Cr(VI) compounds were considered noncarcinogenic at that time, including the chromates and bichromates of hydrogen, lithium, sodium, potassium, rubidium, cesium, and ammonium, and chromic acid anhydride. These Cr(VI) compounds are relatively soluble in water. It was recommended that a 10-hr TWA limit of 25 µg Cr(VI)/m^3 and a 15-minute ceiling limit of 50 µg Cr(VI)/m^3 be applied to these Cr(VI) compounds.

All other Cr(VI) compounds were considered carcinogenic [NIOSH 1975a]. These Cr(VI) compounds are relatively insoluble in water. At that time, NIOSH subscribed to a carcinogen policy that called for "no detectable exposure levels for proven carcinogenic substances" [Fairchild 1976]. The basis for the REL for carcinogenic Cr(VI) compounds, 1 µg Cr(VI)/m^3 TWA, was the quantitative limitation of the analytical method available at that time for measuring workplace exposures to Cr(VI).

NIOSH revised its policy on Cr(VI) compounds in the *NIOSH Testimony on the OSHA Proposed Rule on Air Contaminants* [NIOSH 1988b]. NIOSH testified that although insoluble Cr(VI) compounds had previously been demonstrated to be carcinogenic, there was now sufficient evidence that soluble Cr(VI) compounds were also carcinogenic. NIOSH recommended that all Cr(VI) compounds, whether soluble or insoluble in water, be classified as potential occupational carcinogens based on the OSHA carcinogen policy. NIOSH also recommended the adoption of the most protective of the available standards, the NIOSH RELs. Consequently the REL of 1 µg Cr(VI)/m^3 TWA was adopted by NIOSH for all Cr(VI) compounds.

NIOSH reaffirmed its policy that all Cr(VI) compounds be classified as occupational carcinogens in the *NIOSH Comments on the OSHA Request for Information on Occupational Exposure to Hexavalent Chromium* and the *NIOSH Testimony on the OSHA Proposed Rule on Occupational Exposure to Hexavalent Chromium* [NIOSH 2002, 2005a]. Other NIOSH Cr(VI) policies were reaffirmed or updated at that time [NIOSH 2002, 2005a]. This criteria document updates the NIOSH Cr(VI) policies, including the revised REL, based on its most recent scientific evaluation.

1.3 The REL for Cr(VI) Compounds

NIOSH recommends that airborne exposure to all Cr(VI) compounds be limited to a concentration of 0.2 µg Cr(VI)/m^3 for an 8-hr TWA exposure during a 40-hr workweek. The use of NIOSH Method 7605 (or validated equivalents) is recommended for Cr(VI) determination. The REL represents the upper limit of exposure for each worker during each work shift. Because of the residual risk of lung cancer at the REL, NIOSH further recommends that all reasonable efforts be made to reduce exposures to Cr(VI) compounds below the REL. The available scientific evidence supports the inclusion of all Cr(VI) compounds into this recommendation. The REL is intended to reduce workers' risk of lung cancer associated with occupational exposure to Cr(VI) compounds over a 45-year working lifetime. Although the quantitative analysis is based on lung cancer mortality data, it is expected that reducing airborne workplace exposures will also reduce the nonmalignant respiratory effects of Cr(VI) compounds, which include irritated, ulcerated, or perforated nasal septa.

Workers are exposed to various Cr(VI) compounds in many different industries and workplaces. Currently there are inadequate exposure assessment and health effects data to quantitatively assess the occupational risk of exposure to each Cr(VI) compound in every workplace. NIOSH used the quantitative risk assessment of chromate production workers conducted by Park et al. [2004] as the basis for the derivation of the revised REL for Cr(VI) compounds. This assessment analyzes the data of Gibb et al. [2000b], the most extensive database of workplace Cr(VI) exposure measurements available, including smoking data on most workers. These chromate production workers were exposed primarily to sodium dichromate, a soluble Cr(VI) compound. Although the risk of worker exposure to insoluble Cr(VI) compounds cannot be quantified, the results of animal studies indicate that this risk is likely as great, if not greater than, exposure to soluble Cr(VI) compounds [Levy et al. 1986]. The carcinogenicity of insoluble Cr(VI) compounds has been demonstrated in animal and human studies [NIOSH 1988b]. Animal studies have demonstrated the carcinogenic potential of soluble and insoluble Cr(VI) compounds [NIOSH 1988b, 2002, 2005a; ATSDR 2012]. Recent molecular toxicology studies provide further support for classifying Cr(VI) compounds as occupational carcinogens without providing sufficient data to quantify different RELs for specific compounds [NIOSH 2005a]. Based on its evaluation of the data currently available, NIOSH recommends that the REL apply to all Cr(VI) compounds. There are inadequate data to exclude any single Cr(VI) compound from this recommendation. In addition to limiting airborne concentrations of Cr(VI) compounds, NIOSH recommends that dermal exposure to Cr(VI) be prevented in the workplace to reduce the risk of adverse dermal health effects, including irritation, ulcers, skin sensitization, and allergic contact dermatitis.

2 | Properties, Production, and Potential for Exposure

2.1 Physical and Chemical Properties

Chromium (Cr) is a metallic element that occurs in several valence states, including Cr^{-4} and Cr^{-2} through Cr^{+6}. In nature, chromium exists almost exclusively in the trivalent (Cr[III]) and hexavalent (Cr[VI]) oxidation states. In industry, the oxidation states most commonly found are Cr(0) (metallic or elemental chromium), Cr(III), and Cr(VI).

Chemical and physical properties of select Cr(VI) compounds are listed in Table 2–1. The chemical and physical properties of Cr(VI) compounds relevant to workplace sampling and analysis are discussed further in Chapter 3, "Measurement of Exposure."

2.2 Production and Use in the United States

In the marketplace, the most prevalent materials that contain chromium are chromite ore, chromium chemicals, ferroalloys, and metal. In 2010, the United States consumed about 2% of world chromite ore production in imported materials such as chromite ore, chromium chemicals, chromium ferroalloys, chromium metal, and stainless steel [USGS 2011]. One U.S. company mined chromite ore and one U.S. chemical firm used imported chromite to produce chromium chemicals. Stainless- and heat-resisting-steel producers were the leading consumers of ferrochromium. The United States is a major world producer of chromium metal, chromium chemicals, and stainless steel [USGS 2009]. Table 2–2 lists select statistics of chromium use in the United States [USGS 2011].

Sodium dichromate is the primary chemical from which other Cr(VI) compounds are produced. Currently the United States has only one sodium dichromate production facility. Although production processes may vary, the following is a general description of Cr(VI) compound production. The process begins by roasting chromite ore with soda ash and varying amounts of lime at very high temperatures to form sodium chromate. Impurities are removed through a series of pH adjustments and filtrations. The sodium chromate is acidified with sulfuric acid to form sodium dichromate. Chromic acid can be produced by reacting concentrated sodium dichromate liquor with sulfuric acid. Other Cr(VI) compounds can be produced from sodium dichromate by adjusting the pH and adding other compounds. Solutions of Cr(VI) compounds thus formed can then be crystallized, purified, packaged, and sold. Cr(VI) compounds commonly manufactured include sodium dichromate, sodium chromate, potassium dichromate, potassium chromate, ammonium dichromate, and Cr(VI) oxide. Other materials containing Cr(VI) commonly manufactured include various paint and primer pigments, graphic art supplies, fungicides, and corrosion inhibitors.

Table 2–1. Chemical and physical properties of select hexavalent chromium compounds

Compound	Molecular weight	Boiling point (°C)	Melting point (°C)	Solubility Cold water g/100 cc	°C	Other
Ammonium chromate	152.07	—	Decomposes at 180	40.5	30	Insoluble in alcohol; slightly soluble in NH_3, acetone
Ammonium dichromate	252.06	—	Decomposes at 170	30.8	15	Soluble in alcohol; insoluble in acetone
Barium chromate	253.32	—	—	0.00034	160	Soluble in mineral acid
Calcium chromate (dehydrate)	156.07	—	$-2H_2O$, 200	16.3	20	Soluble in acid, alcohol
Chromium (VI) oxide (chromic acid)	99.99	Decomposes	196	67.45	100	Soluble in alcohol, ether, sulfuric acid, nitric acid
Lead chromate	323.19	Decomposes	844	0.0000058	25	Soluble in acid, alkali; insoluble in acetic acid
Lead chromate oxide	546.39	—	—	Insoluble	—	Soluble in acid, alkali
Potassium chromate	194.19	—	968.3 975	62.9 36	20 20	Insoluble in alcohol
Potassium dichromate	294.18	Decomposes at 500	Triclinic becomes monoclinic at 241.6; Melting point is 398	4.9 102	0 100	Insoluble in alcohol
Silver chromate	331.73	—	Decomposes	0.0014	—	Soluble in NH_4OH, KCN
Sodium chromate	161.97	—	19.92	87.3	30	Slightly soluble in alcohol; soluble in MeOH
Sodium dichromate	261.97	Decomposes at 400 (anhydrous)	—	238 (anhydrous) 180	0 20	Insoluble in alcohol
Strontium chromate	203.61	—	—	0.12	15	Soluble in HCl, HNO_3, acetic acid, NH_4 salts
Zinc chromate	181.36	—	—	Insoluble	Insoluble	Soluble in acid, liquid NH_3; insoluble in acetone

Source: The Merck Index [2006].

Table 2–2. Selected chromium statistics, United States, 2007–2010
(In thousands of metric tons, gross weight)

Statistic	2007	2008	2009	2010
Production, recycling	162	146	141	144
Imports for consumption	485	559	273	499
Exports	291	287	280	274

Source: USGS [2012].

2.3 Potential Sources of Occupational Exposure

2.3.1 Airborne Exposure

Workers are potentially exposed to airborne Cr(VI) compounds in three different workplace scenarios: (1) when Cr(VI) compounds are manufactured from other forms of Cr such as in the production of chromates from chromite ore; (2) when products or substances containing Cr(VI) are used to manufacture other products such as chromate-containing paints; or (3) when products containing other forms of Cr are used in processes and operations that result in the formation of Cr(VI) as a by-product, such as in welding.

Many of the processes and operations with worker exposure to Cr(VI) are those in which products or substances that contain Cr(VI) are used to manufacture other products. Cr(VI) compounds impart critical chemical and physical properties such as hardness and corrosion resistance to manufactured products. Chromate compounds used in the manufacture of paints result in products with superior corrosion resistance. Chromic acid used in electroplating operations results in the deposition of a durable layer of chromium metal onto a base-metal part. Anti-corrosion pigments, paints, and coatings provide durability to materials and products exposed to the weather and other extreme conditions.

Operations and processes in which Cr(VI) is formed as a by-product include those utilizing metals containing metallic chromium, including welding and the thermal cutting of metals; steel mills; and iron and steel foundries. Ferrous metal alloys contain chromium metal in varying compositions, lower concentrations in mild steel and carbon steel, and higher concentrations in stainless steels and other high-chromium alloys. The extremely high temperatures used in these operations and processes result in the oxidation of the metallic forms of chromium to Cr(VI). In welding operations both the base metal of the parts being joined and the consumable metal (welding rod or wire) added to create the joint have varying compositions of chromium. During the welding process, both are heated to the melting point, and a fraction of the melted metal vaporizes. Any vaporized metal that escapes the welding-arc area quickly condenses and oxidizes into welding fume, and an appreciable fraction of the chromium in this fume is in the form of Cr(VI) [EPRI 2009; Fiore 2006; Heung et al. 2007]. The Cr(VI) content of the fume and the resultant potential for Cr(VI) exposures are dependent on several process factors, most importantly the welding process and shield-gas type, and the Cr content of both the consumable material and the base metal [Keane et al. 2009; Heung et al. 2007; EPRI 2009; Meeker et al. 2010].

The bioaccessibility of inhaled Cr(VI) from welding fume may vary depending on the fume-generation source. Characterizations of bioaccessibility and biological indices of Cr(VI) exposure have been reported [Berlinger et al. 2008; Scheepers et al. 2008; Brand et al. 2010].

2.3.2 Dermal Exposure

Dermal exposure to Cr(VI) may occur with any task or process in which there is the potential for splashing, spilling, or other skin contact with material that contains Cr(VI). If not adequately protected, workers' skin may be directly exposed to liquid forms of Cr(VI) as in electroplating baths or solid forms, as in Portland cement. Dermal exposure may also occur because of the contamination of workplace surfaces or equipment. Sanitation and hygiene practices and the use of adequate PPE are important to preventing dermal exposures, contamination of workplace surfaces, and take-home exposures.

2.4 Industries with Potential Exposure

2.4.1 Airborne Exposure

Workers have potential exposures to airborne Cr(VI) compounds in many industries, including chromium metal and chromium metal alloy production and use, electroplating, welding, and the production and use of compounds containing Cr(VI). Primary industries with the majority of occupational exposures to airborne Cr(VI) compounds include welding, painting, electroplating, steel mills, iron and steel foundries, wood preserving, paint and coatings production, chromium catalyst production, plastic colorant producers and users, production of chromates and related chemicals from chromite ore, plating mixture production, printing ink producers, chromium metal production, chromate pigment production, and chromated copper arsenate (CCA) producers [Shaw Environmental 2006]. Industries with limited potential for occupational exposure to Cr(VI) compounds include chromium dioxide, chromium dye, and chromium sulfate production; chemical distribution; textile dyeing; glass production; printing; leather tanning; chromium catalyst use; refractory brick production; woodworking; solid waste incineration; oil and gas well drilling; Portland cement production; non-ferrous superalloy production and use; construction; and makers of concrete products [Shaw Environmental 2006].

2.4.2 Dermal Exposure

The construction industry has the greatest number of workers at risk of dermal exposure to Cr(VI) due to working with Portland cement. Exposures can occur from contact with a variety of construction materials containing Portland cement, including cement, mortar, stucco, and terrazzo. Examples of construction workers with potential exposure to wet cement include bricklayers, cement masons, concrete finishers, construction craft laborers, hod carriers, plasterers, terrazzo workers, and tile setters [CPWR 1999a; NIOSH 2005a; OSHA 2008].

Workers in many other industries are at risk of dermal exposure if there is any splashing, spilling, or other skin contact with material containing Cr(VI). Other industries with reported dermal exposure include chromate production [Gibb et al. 2000a]; electroplating [Makinen and Linnainmaa 2004a]; and grinding of stainless and acid-proof steel [Makinen and Linnainmaa 2004b].

2.5 Number of U.S. Workers Potentially Exposed

The National Occupational Hazard Survey, conducted by NIOSH from 1972 through 1974, estimated that 2.5 million workers were potentially exposed to chromium and its compounds [NIOSH 1974]. It was estimated that 175,000 workers were potentially exposed to Cr(VI) compounds. The National Occupational Exposure Survey (NOES), conducted from 1981 through 1983, estimated that 196,725 workers were potentially exposed to Cr(VI) compounds [NIOSH 1983a]. These estimates are obsolete. They are provided for historical purposes only.

In 1981, Centaur Research, Inc. estimated that 391,400 workers were exposed to Cr(VI) in U.S. workplaces, with 243,700 workers exposed to Cr(VI) only and an additional 147,700 workers exposed to a mixture of Cr(VI) and other forms of chromium [Centaur 1981].

In 1994, Meridian Research, Inc. estimated that the number of production workers in U.S. industries with potential exposure to Cr(VI) was 808,177 [Meridian 1994]. Industries included in the analysis included electroplating, welding, painting, chromate producers, chromate pigment producers, CCA producers, chromium catalyst producers, paint and coatings producers, printing ink producers, plastic colorant producers, plating mixture producers, wood preserving, ferrochromium producers, iron and steel producers, and iron and steel foundries. More than 98 percent of the potentially exposed workforce was found in six industries: electroplating, welding, painting, paint and coatings production, iron and steel production, and iron and steel foundries.

In 2006, OSHA estimated that more than 558,000 U.S. workers were exposed to Cr(VI) compounds [71 Fed. Reg.* 10099 (2006); Shaw Environmental 2006]. The largest number of workers potentially exposed to Cr(VI) were in the following application groups: carbon steel welding (> 141,000), stainless steel welding (> 127,000), painting (> 82,000), electroplating (> 66,000), steel mills (> 39,000), iron and steel foundries (> 30,000), and textile dyeing (> 25,000) [71 Fed. Reg. 10099 (2006); Shaw Environmental 2006]. Within the welding application group (stainless steel and carbon steel combined), the largest numbers of exposed workers were reported in the construction (> 140,000) and general industries (> 105,000). Within the painting application group, the largest number of exposed workers was reported in the general (> 37,000) and construction industries (> 33,000). Table 2–3 summarizes the estimated number of workers exposed by application group [71 Fed. Reg. 10099 (2006)].

In addition to those workers exposed to airborne Cr(VI) compounds, an estimated 1,045,500 U.S. workers are potentially exposed to Cr(VI) in cement [Shaw Environmental 2006]. Most of these workers are exposed to wet cement.

2.6 Measured Exposure in the Workplace

2.6.1 NIOSH Multi-Industry Field Study [Blade et al. 2007]

From 1999 through 2001, NIOSH conducted a Cr(VI) field research study consisting of industrial hygiene and engineering surveys at 21 selected sites representing a variety of industrial sectors, operations, and processes [Blade et al. 2007]. This study characterized workers' exposures to airborne particulate containing Cr(VI) and evaluated existing technologies for controlling these exposures. Evaluation methods included the collection of full work shift, personal breathing-zone (PBZ) air samples for Cr(VI), measurement of ventilation system parameters, and documentation of processes and work practices. Operations and facilities evaluated included chromium electroplating; painting and coating; welding in construction; metal-cutting operations on materials containing chromium in ship breaking; chromate-paint removal with abrasive blasting; atomized alloy-spray coating; foundry operations; printing;

Federal Register. See Fed. Reg. in references.

Table 2–3. Estimated number of workers exposed to Cr(VI) by application group

Application group	Number of exposed workers
Welding (stainless steel and carbon steel)	269,379
Painting	82,253
Electroplating	66,859
Steel mills	39,720
Iron and steel foundries	30,222
Textile dyeing	25,341
Woodworking	14,780
Printing	6,600
Glass producers	5,384
Construction, other*	4,069
Chemical distributors	3,572
Paint and coatings producers	2,569
Solid waste incineration	2,391
Non-ferrous metallurgical uses	2,164
Chromium catalyst users	949
Plastic colorant producers and users	492
Chromium catalyst producers	313
Chromate production	150
Plating mixture producers	118
Printing ink producers	112
Chromium dye producers	104
Refractory brick producers	90
Ferrochromium producers	63
Chromate pigment producers	52
Chromated copper arsenate producers	27
Chromium sulfate producers	11
Total	558,431

Adapted from 71 Fed. Reg. 10099, Table VIII-3 [2006].
*Does not include welding, painting, and woodworking; does include government construction.

and the manufacture of refractory brick, colored glass, prefabricated concrete products, and treated wood products. The field surveys represent a series of case studies rather than a statistically representative characterization of U.S. occupational exposures to Cr(VI). A limitation of this study is that for some operations only one or two samples were collected.

The industrial processes and operations were classified into four categories, using the exposure and exposure-control information collected at each site. Each category was determined based on a qualitative assessment of the relative difficulty of controlling Cr(VI) exposures to the existing REL of 1 µg/m³. The measured exposures were compared with the existing REL.

For exposures exceeding the existing REL, the extent to which the REL was exceeded was considered, and a qualitative assessment of the effectiveness of the existing controls was made. An assessment based on professional judgment determined the relative difficulty of improving control effectiveness to achieve the REL. The four categories into which the processes or operations were categorized are as follows:

1. Those with minimal worker exposures to Cr(VI) in air (Table 2–4).

2. Those with workers' exposures to Cr(VI) in air easier to control to existing NIOSH REL than categories (3) and (4) (Table 2–5).

3. Those with workers' exposures to Cr(VI) in air moderately difficult to control to the existing NIOSH REL (Table 2–6).

4. Those with workers' exposures to Cr(VI) in air most difficult to control to the existing NIOSH REL (Table 2–7).

The results of the field surveys are summarized in Tables 2–4 through 2–7. The results characterize the potential exposures as affected by engineering controls and other environmental factors, but not by the use or disuse of PPE. The PBZ air samples were collected outside any respiratory protection or other PPE (such as welding helmets) worn by the workers. A wide variety of processes and operations were classified as those with minimal worker exposures to Cr(VI) in air (Table 2–4) or where workers' Cr(VI) exposures were determined to be easier to control to the existing REL (Table 2–5). Most of the processes and operations where controlling workers' Cr(VI) exposures to the existing REL were determined to be moderately difficult to control involved joining and cutting metals, when the chromium content of the materials involved was relatively high (Table 2–6). In the category where it was determined to be most difficult to control workers' airborne Cr(VI) exposures to the existing REL, all of the processes and operations involved the application of coatings and finishes (Table 2–7). The classification of these processes, based on the potential relative difficulty of controlling occupational exposures to Cr(VI) in air without reliance on respiratory protection devices, represents qualitative assessments based on the professional judgment of the researchers. Recommendations for reducing workers' exposures to Cr(VI) at these sites are discussed in Chapter 8 and in Blade et al. [2007].

2.6.2 Shaw Environmental Report [2006]

The full-shift exposure data from OSHA and NIOSH site visits, NIOSH industrial hygiene surveys, NIOSH health hazard evaluations (HHEs), OSHA Integrated Management Information System (IMIS) data, U.S. Navy and other government and private sources were compiled to demonstrate the distribution of full-shift personal exposures to Cr(VI) compounds in various industries [Shaw Environmental 2006]. Industry sectors identified as having the majority of occupational exposures include electroplating, welding, painting, production of chromates and related chemicals from chromite ore, chromate pigment production, CCA production, chromium catalyst production, paint and coatings production, printing ink producers, plastic colorant producers and users, plating mixture production, wood preserving, chromium metal production, steel mills, and iron and steel foundries. An estimate of the number of workers exposed to various Cr(VI) exposure levels in each primary industry sector is summarized in Table 2–8 [adapted from Shaw Environmental 2006]. Industry sectors with the greatest number of workers exposed above the revised REL include welding, painting, electroplating, steel mills, and iron and steel foundries. These industries also have the greatest number of workers exposed to Cr(VI) compounds.

Table 2–4. Cr(VI) sampling, category 1 operations (Minimal Cr[VI] exposures).

Operation(s)	SIC[‡] code	NIOSH site no. and description	Key job(s) exposed					Other jobs exposed, full-shift PBZ Cr(VI) exposures in air[†] ($\mu g/m^3$)	Process details, engineering exposure-control measures, other comments
				Full-shift PBZ* Cr(VI) exposures in air[†]					
			Job title	Range, $\mu g/m^3$ (n = no. of values)	Geometric mean, $\mu g/m^3$ (GSD)	Tasks, comments			
Bright chromium electroplating (mfg)	3471	(1) Chromium electroplating and coating processes (mfg)	Production worker	~0.09–0.28 (n = 6)	0.15 (1.6)	Place and remove parts to be plated, tend tanks.	None	No LEV.
Chromium coating processes (non-electroplating) (mfg)	3471	(1) Chromium electroplating and coating processes (mfg)	Production worker	0.27 (n = 1) 0.25 (n = 1)	N/A	Place and remove parts to be coated, tend tanks.	Strip line operator 0.25 $\mu g/m^3$ (n = 1) Dye line operator ~0.10 $\mu g/m^3$ (n = 1)	No LEV. One tank on cad line covered with tarp.
TIG, fusion, dual-shield welding; submerged-arc plasma cutting	3494	(14) Welding and cutting on stainless and mild steels (mfg)	TIG welder	<0.06–<0.08 (n = 6, all ND)	N/A	TIG welding on stainless steel	Fusion, dual-shield weld, submerged-arc plasma cut (all on mild steel); all not detected, <0.2 (n = 15)	Welding fume extractor LEV on welding stations, but contaminant capture poor; none on plasma cutting.
Foundry—casting operations—stainless steel, other ferrous alloys (mfg)	3324	(19) Foundry—stainless steel and other ferrous alloys (mfg)	All casting operations workers	0.008–0.19 (n = 13)	0.032 (2.4)	Melt alloy, pour. Alloy Cr content <0.25%–26%	None	Good LEV in old facility (n = 3 exposure measurements, all ≤0.02), but none yet in new facility.
Stick, MIG welding on steel, galvanized piping and sheet metal (construction)	1711	(20) Welding on piping and sheet metal (construction)	Welder	<0.04–0.42 (n = 7) (n = 4 ND)	N/A	Welding (mainly stick) and grinding, indoors	Welding outdoors, <0.04–0.053 (n = 8) (n = 6 ND)	One indoor area had effective LEV. Other work areas in the open, partially enclosed, or passive ventilation.

See footnotes at end of table.

(Continued)

Table 2–4 (Continued). Cr(VI) sampling, category 1 operations (Minimal Cr[VI] exposures).

Operation(s)	SIC[‡] code	NIOSH site no. and description	Key job(s) exposed					Other jobs exposed, full-shift PBZ Cr(VI) exposures in air[†] (μg/m³)	Process details, engineering exposure-control measures, other comments
				Full-shift PBZ* Cr(VI) exposures in air[†]					
			Job title	Range, μg/m³ (n = no. of values)	Geometric mean, μg/m³ (GSD)	Tasks, comments			
Manufacturing of pre-cast concrete products	3272	(10) Manufacture of pre-cast concrete products	Mixer operator	0.22, 0.36 (n = 2)	N/A	Mixes batches		All other jobs, < 0.02–0.25 (n = 32) (n = 9 ND)	Cr(VI) is natural constituent of Portland cement. Minimal exposure-control measures, no engineering exposure controls.
Foundry—ductile iron (mfg)	3321	(15) Foundry—ductile iron (mfg)	All jobs	< 0.04–0.04 (n = 27) (n = 26 ND)	N/A	All foundry tasks		None	Little to no exposure. LEV in furnace area, but ineffective capture. Elsewhere, general ventilation.
Crushing and recycling of concrete from demolition	1795	(12) Crushing and recycling of concrete from demolition	All jobs	< 0.02–0.03 (n = 4) (n = 3 ND)	N/A	All tasks		None	Cr(VI) is natural constituent of Portland cement. Little to no exposure. Outdoor operations, water-spray dust suppression.
Manufacturing of colored glass products, using chromate pigments	3229	(6) Manufacture of colored glass products	All jobs	< 0.02–0.02 (n = 9) (n = 8 ND)	N/A	All tasks		None	LEV at pigment weighing, and batch weighing and mixing; spray-mist dust suppression at cullet station.

See footnotes at end of table.

(Continued)

Table 2–4 (Continued). Cr(VI) sampling, category 1 operations (Minimal Cr[VI] exposures).

Operation(s)	SIC‡ code	NIOSH site no. and description	Key job(s) exposed				Other jobs exposed, full-shift PBZ Cr(VI) exposures in air† (µg/m³)	Process details, engineering exposure-control measures, other comments
			Job title	Full-shift PBZ* Cr(VI) exposures in air†		Tasks, comments		
				Range, µg/m³ (n = no. of values)	Geometric mean, µg/m³ (GSD)			
Screen printing (mfg) with inks containing chromate pigments	2759	(8) Screen printing (mfg), electronic component mfg	All jobs	<0.02 (n = 4 ND)	N/A	Ink mixing, screen printing	None	No detectable exposure. LEV for ink-mixing, general ventilation with HEPA-filtered supply for screen-printing.
Chromate-conversion treatment process (mfg) for electronic-component boards	3679	(8) Screen printing (mfg), electronic component mfg	All jobs	<0.02 (n = 2 ND)	N/A	Operate chromic-acid tank (chromate conversion)	None	No detectable exposure. LEV for chromic acid tanks, general ventilation for adjacent shipping dept.

Source: Blade et al. [2007].
*Abbreviations: GSD = geometric standard deviation; n = number of samples; LEV = local exhaust ventilation; mfg=manufacturing; MIG = metal inert gas; n = number of samples; ND = not detected; PBZ = personal breathing zone; SIC = Standard Industrial Classification; TIG = tungsten inert gas.
†Concentration value preceded by a "less-than" symbol (<) indicates that the Cr(VI) level in the sampled air was less than the minimum detectable concentration (i.e., the mass of Cr[VI] collected in the sample was less than the analytical limit of detection [LOD]). A concentration value preceded by an "approximately" symbol (~) indicates that Cr(VI) was detectable in the sampled air, but at a level less than the minimum quantifiable concentration (i.e., the mass of Cr[VI] collected in the sample was between the analytical LOD and limit of quantification [LOQ]). These concentration values are less precise than fully quantifiable values.
‡SIC codes were in use when these surveys were conducted. See the SIC Manual at www.osha.gov/pls/imis/sic_manual.html.

Table 2–5. Cr(VI) sampling, category 2 operations (Exposures easier to control to 1 μg Cr[VI]/m³ than category 3 or 4).

Operation(s)	SIC[‡] code	NIOSH site no. and description	Key job(s) exposed					Other jobs exposed, full-shift PBZ Cr(VI) exposures in air[†] (μg/m³)	Process details, engineering exposure-control measures, other comments
			Job title	Full-shift PBZ* Cr(VI) exposures in air[†]					
				Range, μg/m³ (n = no. of values)	Geometric mean, μg/m³ (GSD)	Tasks, comments			
Alodyne/anodize chromium-coating processes (mfg)	3471	(2) Painting and coating processes (mfg)	Chem line operator	0.55, 1.1 (n = 2)	N/A	Tending chromic-acid dip tanks (non-electroplating)		Chemist (lab and waste treatment) 0.82 and 1.2 μg/m³	No LEV. Dip tanks covered with tarps.
TIG welding on stainless steel in sheet-metal fabrication (mfg)	3444	(9) Welding and cutting in sheet-metal fabrication (mfg)	TIG welder	0.65 (n = 1)	N/A	TIG welding on stainless steel		None (welder's exposure inside welding helmet = 0.67 μg/m³)	LEV for welding, but poor capture.
Manufacturing of refractory brick using chromic oxide	3297	(5) Manufacture of refractory brick (non-clay)	Salvage operator	0.04, 1.8 (n = 2)	N/A	Exposure higher when cleaned yellow chromate matl.		All other jobs: 0.012–0.74 (n = 20), GM = 0.052, GSD = 3.4	No LEV on the salvage-material cleaning operation. Local ventilation, and other controls, in other areas.
Manufacturing of chromium sulfate from sodium dichromate	2819	(4) Manufacture of chromium sulfate	Reactor operator	0.22, 1.4 (n = 2)	N/A	Transfer materials, collect process QC samples		Railcar operator. Transfers sodium dichromate solution. 0.12, 0.22 (n = 2)	Reactors equipped with LEV, and anti-frothing surfactant. Railcar unloading is closed process.
Remove chromate-containing paint by abrasive blasting (construction)	1721	(17) Remove paint (by abrasive blast) and reapply (construction)	Painter	0.10–1.3 (n = 8)	0.43 (2.3)	Spot abrasive blasting on steel bridge		Exposures during blowdown and non-chromate repainting tasks, 0.077–0.29 (n = 7)	Work inside containment area for environmental contaminants. Natural ventilation only. Low production job, spot blasting only.

(Continued)

See footnotes at end of table.

Table 2–5. Cr(VI) sampling, category 2 operations (Exposures easier to control to 1 μg Cr[VI]/m³ than category 3 or 4).

Operation(s)	SIC[‡] code	NIOSH site no. and description	Key job(s) exposed					Process details, engineering exposure-control measures, other comments
			Job title	Full-shift PBZ* Cr(VI) exposures in air[†]		Tasks, comments	Other jobs exposed, full-shift PBZ Cr(VI) exposures in air[†] (μg/m³)	
				Range, μg/m³ (n = no. of values)	Geometric mean, μg/m³ (GSD)			
SMAW, FCAW, dual-shield, TIG, MIG welding on stainless, other steels (shipyd)	3731	(16) Welding in shipyard operations	Welder	0.19–0.96 (n = 3)	0.36 (2.4)	SMAW, TIG welding in tight below-deck spaces	TIG, MIG, stick welding in relatively open areas, < 0.04–0.22 (n = 15)	LEV was provided to varying degrees in the tight below-deck spaces by moving flex ducts to work space.
Manufacturing of products from wood treated with Cr-Copper-Arsenate	2452	(11) Manufacture of products from treated wood	Fabricator	Limited evaluation, no full-shift measurements	N/A	Sawing, drilling	None (two short-term samples collected outdoors; no Cr(VI) detected.)	No engineering exposure-control measures used, even indoors. Thus, indoor operations may result in detectable exposures.

Source: Blade et al. [2007].

*Abbreviations: FCAW = flux cored arc welding; GM = geometric mean; GSD = geometric standard deviation; n = number of samples; LEV = local exhaust ventilation; mfg = manufacturing; MIG = metal inert gas; n = number of samples; ND = not detected; PBZ = personal breathing zone; SIC = Standard Industrial Classification; SMAW = shield-metal arc welding; TIG = tungsten inert gas.

[†]Concentration value preceded by a "less-than" symbol (<) indicates that the Cr(VI) level in the sampled air was less than the minimum detectable concentration (i.e., the mass of Cr[VI] collected in the sample was less than the analytical limit of detection [LOD]).

[‡]SIC codes were in use when these surveys were conducted. See the SIC Manual at www.osha.gov/pls/imis/sic_manual.html.

Table 2–6. Cr(VI) sampling, category 3 operations (Exposures moderately difficult to control to 1 µg Cr[VI]/m³)

Operation(s)	SIC[‡] code	NIOSH site no. and description	Key job(s) exposed				Other jobs exposed, full-shift PBZ Cr(VI) exposures in air[†] (µg/m³)	Process details, engineering exposure-control measures, other comments
			Job title	Full-shift PBZ* Cr(VI) exposures in air[†]		Tasks, comments		
				Range, µg/m³ (n = no. of values)	Geometric mean, µg/m³ (GSD)			
Manufacturing of screen-printing inks containing chromate pigments	2893	(3) Manufacture of screen-printing inks	Ink-batch weigher	<0.08–3.0 (n = 4) (n = 1 ND)	0.9 (6.2)	Add pigment (powder), other ingredients, then mix ink batch	Other jobs in process: <0.08–0.4 µg/m³ (n = 6) (n = 4 ND)	LEV (fair) for batch weighing/mixing, and certain other operations. Others only general ventilation.
MIG welding on stainless steel in sheet-metal fabrication (mfg)	3444	(9) Welding and cutting in sheet-metal fabrication (mfg)	MIG Welder	2.8, 5.2 (n = 2)	N/A	MIG welding on stainless steel	None (welder's exposures inside welding helmet = 2.6, 1.0, respectively)	LEV for welding, but poor capture.
MIG, TIG welding, plasma-arc cutting, on stainless-steel sheet metal (mfg)	3444	(9) Welding and cutting in sheet-metal fabrication (mfg)	Welding supervisor	2.0, 3.7 (n = 2)	N/A	MIG, TIG weld, plasma-arc cut, grind, metal forming	None. (Supervisor's exposures inside welding helmet = 8.5, 3.2, respectively)	LEV for welding, but poor capture. Only general ventilation for plasma-arc cutting, no local ventilation.
MIG welding on stainless steel (mfg)	3494	(14) Welding and cutting on stainless and mild steels (mfg)	MIG Welder	0.20–5.5 (n = 4) (n = 1, > 1.0)	0.84 (4.0)	MIG welding (non-automated) on stainless steel	Automated MIG-welder operator (stainless steel) <0.07, <0.08 µg/m³ (n = 2)	Welding fume extractor LEV on welding stations, but contaminant capture poor. Also general ventilation.

See footnotes at end of table.

(Continued)

Table 2–6 (Continued). Cr(VI) sampling, category 3 operations (Exposures moderately difficult to control to 1 μg Cr[VI]/m³)

Operation(s)	SIC[‡] code	NIOSH site no. and description	Key job(s) exposed					Other jobs exposed, full-shift PBZ Cr(VI) exposures in air[†] (μg/m³)	Process details, engineering exposure-control measures, other comments
			Job title	Full-shift PBZ* Cr(VI) exposures in air[†]			Tasks, comments		
				Range, μg/m³ (n = no. of values)	Geometric mean, μg/m³ (GSD)				
Metal cutting (torch and carbon-arc) in ship demolition (shipyard)	4499	(13) Metal cutting in ship demolition (shipyard)	Burner	<0.07–27 (n = 14) (n = 2, >1.0)	0.35 (5.4)		Carbon-arc and torch cutting on steel (some with chromate paint)	Firewatch (assist burner) <0.04–1.0 (n = 10) Supervisor <0.07 (n = 2)	Most work performed outdoors, including a partly enclosed area. Some work indoors, only general ventilation provided there.
Repair welding and cutting on alloy and stainless-steel castings (mfg)	3324	(19) Foundry—stainless steel and other ferrous alloys (mfg)	Welder	0.37–22 (n = 4) (n = 1, <12)	6.6 (7.0)		MIG, TIG, SMAW weld, carbon-arc gouge	None	Welding workload 2- to 3-times normal, on various Cr-content steels and alloys. Cutting on 25% Cr alloy. No local ventilation.

Source: Blade et al. [2007].

*Abbreviations: GM = geometric mean; GSD = geometric standard deviation; n = number of samples; LEV = local exhaust ventilation; mfg = manufacturing; MIG = metal inert gas; n = number of samples; ND = not detected; PBZ = personal breathing zone; SIC = Standard Industrial Classification; TIG = tungsten inert gas.

[†]Concentration value preceded by a "less-than" symbol (<) indicates that the Cr(VI) level in the sampled air was less than the minimum detectable concentration (i.e., the mass of Cr[VI] collected in the sample was less than the analytical limit of detection [LOD]). For some other samples in these sets, Cr(VI) was detectable in the sampled air but at a level less than the minimum quantifiable concentration (i.e., the mass of Cr[VI] collected in the sample was between the analytical LOD and limit of quantification [LOQ]). These concentration values are less precise than fully quantifiable values.

[‡]SIC codes were in use when these surveys were conducted. See the SIC Manual at www.osha.gov/pls/imis/sic_manual.html.

Table 2–7. Cr(VI) Sampling, Category 4 Operations (Exposures Most Difficult to Control to 1 μg Cr[VI]/m³).

Operation(s)	SIC[‡] code	NIOSH site no. and description	Key job(s) exposed				Other jobs exposed, full-shift PBZ Cr(VI) exposures in air[†] (μg/m³)	Process details, engineering exposure-control measures, other comments
			Job title(s)	Full-shift PBZ* Cr(VI) exposures in air[†]		Tasks, comments		
				Range, μg/m³ (n = no. of values)	Geometric mean, μg/m³ (GSD)			
Spray application and re-sanding of chromate-containing paints (mfg)	3479	(2) Painting and coating processes (mfg)	Painter	3.8–55 (n = 5)	16 (3.4)	Spray/sand/clean up. Paints: 1–30% chromates	Painter's helpers (same work areas) 2.4–22 (n = 4)	Painting in fully and partially enclosed paint booths—effectiveness judged as fair.
Spray application and re-sanding of chromate-containing paints (mfg)	3728	(7) Painting and associated re-sanding (mfg)	Painter	< 0.02–4.3 (n = 13)	0.23 (6.3)	Spraying paint, some sanding. Paints: 1–30% chromates	Assemblers using rotary-disc sanders 0.27–2.1 (n = 4)	Fully enclosed paint booths. Vacuum-attached disc sanders. Both judged as fair. Other workers' exposures were lower.
Hard chromium electroplating (mfg)	3471	(1) Chromium electroplating and coating processes (mfg)	Plater	3.0–16 (n = 4)	7.9 (2.0)	Place and remove parts to be plated, tend tanks.	Lab tech 9.0 μg/m³ when adding CrO₃ flake. Otherwise, lab workers 0.22, 0.27 (n = 3)	Mist suppressant, push-pull LEV, tarps used on tanks. Lab workers work at tanks along with lab duties.
Hard and bright chromium electroplating (mfg)	3471	(18) Chromium electroplating (mfg)	Plater	0.22–8.3 (n = 12)	2.5 (2.6)	Place and remove parts to be plated, tend tanks.	None	Platers work throughout plant, various plating tanks. LEV on all tanks, new mist suppressant on one.

(Continued)

See footnotes at end of table.

Table 2–7 (Continued). Cr[VI] Sampling, Category 4 Operations (Exposures Most Difficult to Control to 1 μg Cr[VI]/m³).

Operation(s)	SIC[‡] code	NIOSH site no. and description	Key job(s) exposed					Other jobs exposed, full-shift PBZ Cr(VI) exposures in air[†] (μg/m³)	Process details, engineering exposure-control measures, other comments
			Job title(s)	Full-shift PBZ* Cr(VI) exposures in air[†]			Tasks, comments		
				Range, μg/m³ (n = no. of values)	Geometric mean, μg/m³ (GSD)				
Atomized Cr-alloy spray-coating operation (industrial maintenance)	1799	(21) Cr-alloy metalization coating operation (industrial maintenance)	Production worker	≥ 820, ≥ 1900 (n = 2)	N/A		Prep surfaces by abrasive blasting. Then spray coating.	Supervisor, entered enclosed work area: 330 Other supervisors: 44, 47 Abrasive-pot tender: 7.0	Work area inside large boiler, resurfacing heat-exchange tubes. Electric arc melts alloy, then compressed air propels to surface.

Source: Blade et al. [2007].

*Abbreviations: GM = geometric mean; GSD = geometric standard deviation; n = number of samples; LEV = local exhaust ventilation; mfg = manufacturing; n = number of samples; ND = not detected; PBZ = personal breathing zone; SIC = Standard Industrial Classification.

[†]Concentration value preceded by a "less-than" symbol (<) indicates that the Cr(VI) concentration in the sampled air was less than the minimum detectable concentration (i.e., the mass of Cr[VI] collected in the sample was less than the analytical limit of detection [LOD]). For some other samples in these sets, Cr(VI) was detectable in the sampled air, but at a level less than the minimum quantifiable concentration (i.e., the mass of Cr[VI] collected in the sample was between the analytical LOD and limit of quantification [LOQ]). These concentration values are less precise than fully quantifiable values. Additionally, a concentration value preceded by a "greater-than-or-equal-to" symbol (≥) indicates that the reported value is an estimate, and the "true" concentration likely is greater, because of air-sampling pump failure before the end of the intended sampling period.

[‡]SIC codes were in use when these surveys were conducted. See the SIC Manual at www.osha.gov/pls/imis/sic_manual.html.

Table 2–8. Full-shift 8-Hour TWA personal Cr(VI) exposures in primary industry sectors

Industry	Total no. exposed workers	Below LOD	LOD to 0.25 µg/m³	0.25 to 0.5 µg/m³	0.5 to 1 µg/m³	≥ 1 µg/m³
Welding	247,269	47,361	12,588	50,709	75,722	77,307
Painting	82,254	11,283	20,120	17,766	12,876	20,209
Electroplating	66,857	0	21,410	27,470	2,028	16,149
Steel mills	39,720	10,038	9,390	6,417	8,456	5,419
Iron and steel foundries	30,222	4,184	11,875	3,481	4,578	6,104
Paint and coating production	2569	400	1443	38	38	650
Plastic colorant producers; users	492	37	15	15	0	425
Chromium catalyst production	313	0	127	25	31	130
Chromate chemical production	150	1	89	24	24	12
Plating mixture producers	118	0	16	80	0	22
Printing ink production	112	27	4	3	17	61
Chromium metal producers	63	16	8	9	17	13
Chromate pigment production	52	0	0	0	1	51
CCA production	27	0	12	0	5	10

Source: Shaw Environmental [2006].
Abbreviations: CCA = chromated copper arsenate; LOD = limit of detection; TWA = time-weighted average.

Industry sectors that were identified with a lesser potential for airborne Cr(VI) exposure include chromium dioxide, chromium dye, and chromium sulfate production; chemical distribution; textile dyeing; glass production; printing; leather tanning; chromium catalyst use; refractory brick production; woodworking; solid waste incineration; oil and gas well drilling; Portland cement production; non-ferrous superalloy production and use; construction; and makers of concrete products [Shaw Environmental 2006].

More detailed, extensive exposure data by industry sector, process type, and job or operation description are available in the Shaw Environmental Report [2006] and the OSHA Final Rule on Hexavalent Chromium [71 Fed. Reg. 10099 (2006)].

2.6.3 Welding and Thermal Cutting of Metals

Welders are the largest group of workers potentially exposed to Cr(VI) compounds. Cr(VI)

exposures to welders are dependent on several process factors, most importantly the welding process and shield-gas type, and the Cr content of both the consumable material and the base metal [Keane et al. 2009; Heung et al. 2007; EPRI 2009; Meeker et al. 2010]. The exposure data associated with different welding processes has been reported [Shaw Environmental 2006; 71 Fed. Reg. 10099 (2006)].

The Electric Power Research Institute (EPRI) analyzed Cr(VI) exposures during welding and thermal cutting activities conducted at electric utility operations [EPRI 2009]. EPRI reported Cr(VI) exposures associated with shielded-metal arc welding (SMAW) and gas-metal arc welding (GMAW), with geometric means (GMs) of 1.4 $\mu g/m^3$ for SMAW and 1.3 $\mu g/m^3$ for GMAW. This was higher than gas-tungsten arc welding (GTAW), with a GM exposure of 0.14 $\mu g/m^3$. All exposure values represent full work shift TWA personal-breathing-zone exposures to particulate-borne Cr(VI) in air. Metal cutting with arc gouging resulted in GM exposures of 12 $\mu g/m^3$. All exposures were below the OSHA PEL when GTAW was used. During the use of SMAW, exposures were below the PEL only when both the consumable metal contained less than 3% Cr and adequate ventilation was provided [EPRI 2009].

Others have found similar associations between Cr(VI) exposures and welding processes, materials, and ventilation. Meeker et al. [2010] reported GM exposures of about 5 $\mu g/m^3$ for SMAW and 0.7–0.8 $\mu g/m^3$ for GTAW, versus a maximum exposure of 0.4 $\mu g/m^3$ for GTAW, with many exposures below detection limits. In addition to welding process type, other predictors of Cr(VI) inhalation exposures included the base metal Cr content and whether local exhaust ventilation (LEV) was used.

The environmental conditions of the work site also affect worker Cr(VI) exposures. For example, when welding outdoors the worker's exposure level depends on how strong the wind is, what direction the wind is moving, where the worker is standing in relation to the welding plume, and where the LEV is positioned [NIOSH 1997].

2.7 Occupational Exposure Limits

The NIOSH REL for all Cr(VI) compounds is 0.2 μg Cr(VI)/m^3 8-hr TWA. Values for other U.S. occupational exposure limits (OELs) are also listed in Table 2–9. Values for OELs from other countries are presented in Table 2–10.

2.8 IDLH Value

An immediately dangerous to life or health (IDLH) condition is one that poses a threat of exposure to airborne contaminants when that exposure is likely to cause death or immediate or delayed permanent adverse health effects or prevent escape from such an environment [NIOSH 2004]. The purpose of establishing an IDLH value is (1) to ensure that the worker can escape from a given contaminated environment in the event of failure of the respiratory protection equipment and (2) is considered a maximum level above which only a highly reliable breathing apparatus providing maximum worker protection is permitted [NIOSH 2004]. The IDLH for chromic acid and chromates is 15 mg Cr(VI)/m^3 [NIOSH 1994a].

2.9 Future Trends

Industry sectors with the greatest number of workers exposed to Cr(VI) compounds, and the largest number of workers exposed to Cr(VI) compounds above the revised REL, include welding, painting, electroplating, steel mills, and iron and steel foundries [Shaw Environmental 2006; 71 Fed. Reg. 10099 (2006)]. Recent national and international regulations on workplace and environmental Cr(VI) exposures have stimulated the research and development

Table 2–9. U.S. occupational exposure limits for Cr(VI) compounds*

Agency	OEL	Cr(VI) compound(s)	8-hr TWA µg Cr(VI)/m³
NIOSH	REL	All	0.2
OSHA	PEL		5
ACGIH	TLV	Water-soluble	50
		Insoluble	10

Source: ACGIH [2011a]; OSHA [2007].
*Specific Cr(VI) compounds such as calcium, lead, and strontium chromate may have distinct OELs.

Table 2–10. Occupational exposure limits for Cr(VI) compounds in various countries*

Country	Insoluble Cr(VI) TWA (µg/m³)	Soluble Cr(VI) TWA (µg/m³)
Australia	50	50
Canada	10	50
Alberta	10	20
British Columbia	10	50
Quebec		
Hong Kong	10	50
Ireland	50	50
Japan	10	10
Mexico	10	50
Netherlands	10	25
Poland	100	100
Sweden	20	20
United Kingdom	50	50

Source: ACGIH [2011b].
*Specific Cr(VI) compounds such as calcium, lead, strontium, and zinc chromate may have distinct OELs.

of substitutes for Cr(VI). Some defense-related industries are eliminating or limiting Cr(VI) use where proven substitutes are available [76 Fed. Reg. 25569 (2011)]. However, it is expected that worker exposure to Cr(VI) compounds will continue in many operations until acceptable substitutes have been developed and adopted. It is expected that some existing exposures, such as Cr(VI) exposure during the removal of lead chromate paints, will continue to be a risk of Cr(VI) exposure to workers for many years [71 Fed. Reg. 10099 (2006)].

Some industries, such as woodworking, printing ink manufacturing, and printing have decreased their use of Cr(VI) compounds [71 Fed. Reg. 10099 (2006)]. However, many of these workplaces have only a small number of employees or low exposure levels.

3 Measurement of Exposure

Recently developed analytical methods provide an improved ability to determine hexavalent chromium (Cr[VI]) concentrations in workplace air. These methods and sampling considerations for Cr(VI) compounds have been reviewed [Ashley et al. 2003]. NIOSH Method 7605 for Cr(VI) determination in the laboratory and NIOSH Method 7703 for Cr(VI) determination in the field are published in the *NIOSH Manual of Analytical Methods* [http://www.cdc.gov/niosh/nmam] [NIOSH 1994b]. These methods provide improved Cr(VI) measurement by allowing for the detection of Cr(VI) (versus total chromium), quantification of Cr(VI) at trace levels, and measurement of Cr(VI) in soluble and insoluble chromate compounds. NIOSH Method 7605, OSHA Method ID-215, and international consensus standard analytical methods can accurately quantify workplaces exposures at the recommended exposure limit (REL) of 0.2 µg Cr(VI)/m^3 8-hr TWA [NIOSH 2003b; OSHA 2006; ASTM 2002; ISO 2005].

3.1 Air-Sampling Methods

3.1.1 Air Sample Collection

There are several methods developed by NIOSH and others to quantify Cr(VI) levels in workplace air. Specific air-sampling procedures such as sampling airflow rates and recommended sample-air volumes are specified in each of the methods, but they share similar sample-collection principles. All are suitable for the collection of long-term, time-integrated samples to characterize time-weighted average (TWA), personal breathing zone (PBZ) exposures across full work shifts.

Each PBZ sample is collected using a battery-powered air-sampling pump to draw air at a measured rate through a plastic cassette containing a filter selected in accordance with the specific sampling method and the considerations described above. The airflow rate of each air-sampling pump should be calibrated before and after each work shift it is used, and the flow rate adjusted as needed according to the nominal rate specified in the method. Usually when measuring a PBZ exposure, the air inlet of the filter cassette is held in the worker's breathing zone by clipping the cassette to the worker's shirt collar, and the pump is clipped to the worker's belt, with flexible plastic tubing connecting the filter cassette and pump. The air inlet should be located so that the exposure is measured outside a respirator or any other personal protective equipment (PPE) being worn by the worker.

When sampling for welding fumes, the filter cassette should be placed inside the welding helmet to obtain an accurate measurement of the worker's exposure [OSHA 1999b; ISO 2001]. The practice of placing the sampling device inside PPE applies only to PPE that is not intended to provide respiratory protection, such as welding helmets or face shields. If the PPE has supplied air, such as a welding hood or an abrasive blasting hood, then the sample is taken outside the PPE [OSHA 1999b].

For the collection of an area air sample, an entire sampling apparatus (pump, tubing, filter cassette) can be placed in a stationary location. This method can also be used to collect short-term task samples (e.g., 15 minutes), if high enough concentrations are expected, so that the much smaller air volume collected during the short sample duration contains enough Cr(VI) to exceed the detection limits.

The procedures specified in each method for handling the samples and preparing them for on-site analysis or shipment to an analytical laboratory should be carefully followed, including providing the proper numbers of field-blank and media-blank samples specified in the method.

3.1.2 Air Sampling Considerations

Important sampling considerations when determining Cr(VI) levels in workplace air have been reviewed [Ashley et al. 2003]. One of the most important considerations is the reduction of Cr(VI) to Cr(III) during sampling and sample preparation. Another concern is the possibility of oxidation of Cr(III) to Cr(VI) during sample preparation. Factors that affect the reduction of Cr(VI) or oxidation of Cr(III) include the presence of other compounds in the sampled workplace air, which may affect reduction or oxidation (notably iron, especially Fe[II]), the ratio of Cr(VI) to Cr(III) concentrations in the sample, and solution pH [Ashley et al. 2003]. The pH of a solution is an important factor, because in acidic conditions the reduction of Cr(VI) is favorable, while in basic conditions Cr(VI) is stabilized. The sampling and analytical methods developed for the determination of Cr(VI) in the workplace attempt to minimize the influence of these redox reactions in order to obtain accurate Cr(VI) measurements. Using NIOSH Method 7703 in the field is one option to minimize the reduction of Cr(VI) that may occur during sample transport and storage [Marlow et al. 2000; Wang et al. 1999].

It is important to select a filter material that does not react with Cr(VI). All filter types to be used for sampling should be tested before use, but ordinarily polyvinyl chloride (PVC) filters are recommended (NIOSH Method 7605; OSHA Method ID-215). Other suitable filter materials that are generally acceptable for airborne Cr(VI) sampling include polyvinyl fluoride (PVF), polytetrafluoroethylene (PTFE), PVC- and PVF-acrylic copolymers, and quartz fiber filters [Ashley et al. 2003].

3.2 Analytical Methods

3.2.1 Cr(VI) Detection in Workplace Air

There are several methods developed by NIOSH and others to quantify Cr(VI) levels in workplace air. NIOSH Method 7605 describes the determination of Cr(VI) levels in workplace air by ion chromatography [NIOSH 2003b]. This method is a modification of previous NIOSH Methods 7604 and 7600, employing the hot plate extraction and ion chromatographic separation method of the former and the spectrophotometric detection technique of the latter. NIOSH Method 7605 also includes ultrasonic extraction as an optional sample preparation method for Cr(VI) [Wang et al. 1999]. The limits of detection (LOD) per sample is 0.02 µg for NIOSH Method 7605. OSHA Method ID-215 also uses ion chromatography to separate Cr(VI) [OSHA 1998, 2006]. The OSHA method employs a precipitation reagent to prevent Cr(III) oxidation to Cr(VI) during sample preparation, while NIOSH Method 7605 relies on sonication and/or a nitrogen atmosphere to achieve the same end.

NIOSH Method 7703 measures Cr(VI) levels by field-portable spectrophotometry [NIOSH 2003a]. This method is designed to be used in the field with portable laboratory equipment but can also be used in the fixed-site laboratory. It is a relatively simple, fast, and sensitive method for Cr(VI) determination [Wang et al. 1999; Marlow et al. 2000]. The method uses ultrasonic extraction instead of hotplate extraction, and solid-phase extraction instead of ion chromatography to isolate Cr(VI). Its estimated LOD is 0.08 µg per sample. The method has been modified to enable the determination of insoluble Cr(VI) compounds [Hazelwood et al. 2004]. Sequential extraction of soluble and insoluble Cr(VI) compounds in air filter samples, as described in ISO 16740, has been evaluated [Ashley et al. 2009].

Boiano et al. [2000] conducted a field study to compare results of airborne Cr(VI) determination obtained using NIOSH Methods 7605 and 7703 and OSHA Method ID-215 (Table 3–1). All three of these methods use extraction of the PVC filter in alkaline buffer solution, chemical isolation of Cr(VI), complexation of Cr(VI) with 1,5-diphenylcarbazide, and spectrometric measurement. However, there are specific differences regarding sample handling in each method. Three sets of 20 side-by-side air samples (10 at each facility on each of three sampling media) were collected at a chromic acid electroplating operation and a spray paint operation, and analyzed using the three methods. No statistically significant differences were found between the mean Cr(VI) values obtained using the three methods ($P < 0.05$). Overall results obtained using NIOSH Method 7703 were slightly higher (statistically significant) than those obtained using OSHA ID-215.

International standards for the determination of Cr(VI) in workplace air are available that can accurately quantify Cr(VI) exposures at the REL. American Society for Testing and Materials (ASTM) Method D6832-02, "*Standard Test Method for the Determination of Hexavalent Chromium in Workplace Air by Ion Chromatography and Spectrophotometric Measurement Using 1,5-Diphenylcarbazide*," allows for the determination of airborne Cr(VI) [ASTM 2002]. International Organization for Standardization (ISO) 16740, "*Workplace Air—Determination of Hexavalent Chromium in Airborne Particulate Matter—Method by Ion Chromatography and Spectrophotometric Measurement using Diphenylcarbazide*," provides a method to extract Cr(VI) compounds of different solubilities [ISO 2005]. Sulfate buffers are suitable for extraction of Cr(VI) from soluble and sparingly soluble compounds, while carbonate buffers are required for the dissolution of Cr(VI) from insoluble chromate compounds [Hazelwood et al. 2004]. Several other validated procedures for the sampling and analysis of Cr(VI) in occupational settings have been published in the United Kingdom, France, and Germany [Ashley et al. 2003].

3.2.2 Wipe Sampling Methods

NIOSH, OSHA, and ASTM have developed methods that can be used to detect Cr(VI) by using wipe samples. Information about surface and dermal sampling is available [ASTM 2011]. However, there currently are no consensus criteria for interpreting wipe sampling data for chromium. Analytical results from wipe sampling and analysis should be viewed as qualitative or semi-quantitative. OSHA Method W-4001 is a wipe method specific for Cr(VI) sampling [OSHA 2001]. NIOSH Method 9102, "*Elements on Wipes*," is a simultaneous elemental analysis that is not compound specific [NIOSH 2003d]. ASTM D6966, "*Standard Practice for the Collection of Dust Samples Using Wipe Sampling Methods for Subsequent Determination of Metals*" [ASTM 2003] applies to metals determination, so the same sampling procedure can be applicable to the collection of Cr(VI) in surface dust. Sample preparation and analysis procedures using this method for Cr(VI) determination would be similar to those for the airborne Cr(VI) methods in Section 3.2.1. However, media and matrix effects could be problematic for the reasons already discussed (i.e., biases in Cr(VI) measurement due to redox reactions with the sampling media and/or the co-sampled matrix).

NIOSH Method 9101, "*Hexavalent Chromium in Settled Dust Samples,*" allows for screening of soluble Cr(VI) in settled dust [NIOSH 1996a]. Estimation of Cr(VI) in dust can be obtained by laboratory analysis for Cr(VI) using NIOSH Method 7605 or equivalent methods.

Table 3–1. Comparison of NIOSH and OSHA analytical methods for airborne hexavalent chromium determination

Parameter	NIOSH 7605	OSHA ID-215	NIOSH 7703
Sample collection, handling and storage:			
Media	PVC 37 mm; 5.0 µm Cellulose backup pad	PVC 37 mm; 5.0 µm Cellulose backup pad	PVE, MCE, or PTFE 37 mm; 5.0, 0.8, 1.0 µm Cellulose backup pad
Equipment	Personal sampling pump	Personal sampling pump	Personal sampling pump
Flow rate	1–4 L min^{-1}	2 L min^{-1}	1–4 L min^{-1}
Sample preparation for shipment to laboratory	Using Teflon®-coated tweezers, transfer filter to 20 mL glass vial with Teflon cap liner	Using Teflon-coated tweezers, transfer filter to 20 mL glass vial with Teflon cap liner	Not applicable if analyzed on-site. Same sample handling as NIOSH 7605 and OSHA ID-215 if analyzed off-site.
Sample refrigeration	Optional	4 °C	None required
Sample preparation and analysis:			
Extraction solution	2% NaOH/3% Na$_2$CO$_3$ or 0.05 M (NH$_4$)$_2$SO$_4$/0.05 M NH$_4$OH (pH 8)	10% Na$_2$CO$_3$/2% NaHCO$_3$/ phosphate buffer/Mg II (as MgSO$_4$) (pH 8)	0.05 M (NH$_4$)$_2$SO$_4$/0.05 M NH$_4$OH (pH 8)
Extraction equipment	Hot plate	Hot plate	Ultrasonic bath
Cr(VI) isolation	Ion chromatography	Ion chromatography	Strong anion exchange solid phase extraction
Eluent	0.25 M (NH$_4$)$_2$SO$_4$/ 0.1M NH$_4$OH	0.25 M (NH$_4$)$_2$SO$_4$/ 0.1M NH$_4$OH	0.5M (NH$_4$)$_2$SO$_4$/ 0.1M NH$_4$OH
Post-column reagent (derivatization)	2 mM 1,5 diphenyl-carbazide/10% methanol/1 M H$_2$SO$_4$	2 mM 1,5 diphenyl-carbazide/10% methanol/1 M H$_2$SO$_4$	1,5 diphenylcarbazide/acetonitrile solution added to eluent acidified with 1 M HCl
Analyte	Cr-DPC complex	Cr-DPC complex	Cr-DPC complex
Detection	UV-Vis: 540 nm	UV-Vis: 540 nm	UV-Vis: 540 nm
LOD/LOQ/µg	0.02/0.06	0.01/0.03	0.09/0.27
Accuracy	±16.5%	±12.9%	±16.8%

Source: Boiano et al. [2000].
Abbreviations: DPC = diphenylcarbazide/diphenylcarbazone; LOD/LOQ = limit of detection/limit of quantitation; MCE = mixed cellulose ester; PTFE = polytetrafluoroethylene; PVC = polyvinylchloride; UV-Vis = ultraviolet-visible.

3.3 Biological Markers

Biological markers, or biomarkers, can serve several purposes where there is epidemiological evidence that exposure causes a particular disease: answering questions of intensity and timing of exposure; testing the effectiveness of controls; assessing subgroups within a worker population; and functioning as an indicator of early disease [Schulte 1995]. Research is ongoing to identify reliable quantifiable biomarkers of Cr(VI) occupational exposure that can indicate exposure levels, effects of exposure, or early disease conditions. The biological markers of Cr(VI) exposure and effect have been reviewed [ATSDR 2012]. Biomarkers should be evaluated carefully as variables, including diet, capacity to reduce Cr(VI), type of occupational exposure, sensitivity of the analytical method used, and other factors affect results. Biomarkers for Cr(VI) compounds are currently of uncertain value as early indicators of potential health effects related to Cr(VI) exposure [NIOSH 2005a].

An important consideration in biological testing for Cr(VI) is the reduction of Cr(VI) to Cr(III) throughout the body. Some biological markers distinguish Cr(VI) levels while others assess only total chromium levels because of the varying distribution of Cr(III) and Cr(VI) within body compartments. Inhalation is the primary route of concern for occupational Cr(VI) exposure. Inhaled Cr(VI) enters the respiratory system, where it may remain, be reduced, or enter the bloodstream. Cr(VI) may be reduced to Cr(III) in the lungs or plasma and excreted as Cr(III) in the urine. Cr(VI) that is not reduced in the plasma may enter erythrocytes and lymphocytes. This distribution of absorbed Cr(VI) permits the biological monitoring of Cr in urine, whole blood, plasma, and blood cells in workers exposed to Cr(VI) [Miksche and Lewalter 1997].

Urinary chromium levels have been extensively studied. They are a measure of total chromium exposure as Cr(VI) is reduced within the body to Cr(III). Blood Cr levels are lower than urinary levels. Biological monitoring of blood chromium requires careful techniques and equipment to avoid contamination of the samples, and such monitoring requires a sensitive method of analytical detection. Measurement of erythrocyte Cr levels is a measure of Cr(VI) exposure, because Cr(VI) passes through the cell membranes, but Cr(III) does not [Gray and Sterling 1950].

3.3.1 Biological Markers of Exposure

3.3.1.1 Measurement of chromium in urine

Urinary chromium levels are a measure of total chromium exposure as Cr(VI) is reduced within the body to Cr(III). The ACGIH has Biological Exposure Indices (BEIs) for Cr(VI) compounds as a water-soluble fume [ACGIH 2011a]. The BEI for total chromium in urine measured at the end of the shift at the end of the workweek is 25 µg/L. The BEI for the increase in total chromium during a shift is 10 µg/L.

Gylseth et al. [1977] reported a significant correlation ($P < 0.001$) between workplace Cr exposure and urinary Cr concentration after work in five alloyed steel welders. It was assumed that most of their exposure was to soluble Cr(VI). A urinary Cr concentration of 40–50 µg Cr per liter of urine corresponded to an approximate workplace exposure of 50 µg Cr/m^3.

Lindberg and Vesterberg [1983] measured with personal air samplers the Cr(VI) exposures of eight chrome platers and monitored their urinary Cr concentrations. The urinary Cr levels increased from Monday morning until

Tuesday afternoon and then remained constant throughout the workweek. The Monday and Thursday preshift and postshift urinary Cr level and exposure were also monitored on a larger group of 90 chrome platers. Exposure correlated with Thursday postshift urinary Cr levels with exposures of approximately 2 µg/m^3, correlating with ≤ 100 nmol Cr/l urine.

Angerer et al. [1987] measured Cr concentrations in the erythrocytes, plasma, and urine of 103 MMA welding and/or metal inert gas (MIG) welders. Personal air monitoring was also conducted; chromium trioxide exposures ranged from < 1 to 50 µg/m^3. The urinary Cr concentrations ranged from 5.40 to 229.4 µg/l; approximately 5 and 200 times higher than the level of non-exposed people. Erythrocyte, plasma, and urine Cr levels were highly correlated ($P < 0.0001$). The authors reported that plasma chromium levels of approximately 10 µg/l and urine Cr levels of 40 µg/l corresponded to an external exposure of 100 µg CrO_3/m^3, while erythrocyte chromium concentrations greater than 0.60 µg/l indicated exposures greater than 100 µg CrO_3/m^3.

Minoia and Cavalleri [1988] measured urinary Cr levels in dichromate production workers exposed predominantly to Cr(VI) or Cr(III). A correlation was found between Cr(VI) exposure as measured by personal air sampling and postshift urinary levels. Cr(VI) was not detected in the urine samples, indicating the in vivo reduction of Cr(VI) to Cr(III).

Liu et al. [1998] reported a correlation between air and urinary chromium concentrations in hard-chrome platers, nickel-chrome electroplaters, and aluminum anode-oxidation plant workers. Hard-chrome plating workers had the highest air and urinary chromium concentrations, with geometric means of 4.2 µg Cr/m^3 TWA for air and 2.44 µg/g creatinine for urine.

Chen et al. [2008] reported an association of skin disease and/or smoking habit with elevated urinary Cr levels in cement workers. Smoking increased urinary Cr levels an average of 25 µg/L; hand eczema increased urinary Cr levels more than 30 µg/L. Workers with skin disease who were also smokers had higher urinary Cr levels than those with only skin disease or smoking habits. Workers who did not regularly wear PPE had higher average urinary Cr levels, with the difference between glove users and non-users being the greatest ($P < 0.001$).

Individual differences in the ability to reduce Cr(VI) have been demonstrated [Miksche and Lewalter 1997]. Individuals with a weaker capacity to reduce Cr(VI) have lower urine Cr levels compared with individuals who have a stronger capacity to reduce Cr(VI). Therefore, analyzing only urinary Cr levels may not provide an accurate analysis of occupational exposure and health hazard.

3.3.1.2 Measurement of chromium in blood, plasma, and blood cells

Plasma or whole blood chromium levels are indicative of total chromium exposure because Cr(VI) may be reduced to Cr(III) in the plasma. Intracellular chromium levels are indicative of Cr(VI) exposure because Cr(VI) passes through cell membranes, while Cr(III) does not [Gray and Sterling 1950]. The chromium concentration inside erythrocytes indicates exposure to Cr(VI) sometime during the approximate 120-day lifespan of the cells. There are two advantages to the monitoring of chromium levels in erythrocytes (red blood cells) versus urine: (1) the sampling time may be relatively independent of the time of exposure, and (2) it permits the determination of Cr(VI), rather than only total chromium, absorption [Wiegand et al. 1988].

Wiegand et al. [1985] investigated the kinetics of ^{51}Cr(VI) uptake into human erythrocytes in vitro. Two different first-order processes, with half-life times of 22.7 seconds and 10.4 minutes, were observed when erythrocytes were incubated with sodium dichromate concentrations ranging from 10 µM to 50 mM. Approximately 15 percent of the administered dose of Cr(VI) remained in the plasma after a 2-hour incubation. The maximal capacity for Cr(VI) uptake into erythrocytes was 3.1×10^8 chromate ions per cell, per minute.

Many variables can affect chromium levels in the blood, including diet, individual capacity to reduce Cr(VI), and type of occupational exposure. Corbett et al. [1998] reported an enhanced in vitro ability to reduce Cr(VI) in the plasma from an individual who had recently eaten, in comparison with a fasted individual. A concentration-dependent distribution of Cr between the erythrocytes (RBCs) and plasma was reported. A higher Cr(VI) concentration was associated with a higher Cr(VI) concentration in erythrocytes, resulting in a lower plasma to erythrocyte ratio of total chromium.

Individual differences in the ability to reduce Cr(VI) have been demonstrated [Miksche and Lewalter 1997]. Individuals with a weaker plasma capacity to reduce Cr(VI) have elevated plasma Cr(VI) levels in comparison with individuals who have a stronger capacity to reduce Cr(VI). Therefore, elevated blood plasma levels may be indicative of high chromium exposures and/or a low plasma ability to reduce Cr(VI).

Cr(VI) uptake into erythrocytes may also be dependent on the Cr(VI) particle size [Miksche and Lewalter 1997]. Smaller particles, as in welding fume exposure (< 0.5 µm), may be more efficiently reduced in the lungs than larger particles, such as those of chromate dust exposure (> 10 µm).

Minoai and Cavalleri [1988] measured serum and erythrocyte Cr levels in dichromate production workers exposed predominantly to Cr(VI) compounds (chromic trioxide or potassium dichromate) or Cr(III) (basic chromium sulphate) compounds. Workers exposed predominantly to Cr(VI) compounds had lower serum and higher erythrocyte Cr levels compared with predominantly Cr(III)-exposed workers, providing evidence of an enhanced ability of Cr(VI) to enter erythrocytes compared with Cr(III).

Angerer et al. [1987] measured Cr concentrations in the erythrocytes, plasma, and urine of 103 MMA and/or MIG welders. Personal air monitoring was also conducted. Airborne chromium trioxide concentrations for MMA welders ranged from < 1 to 50 µg/m^3, with 50% < 4 µg/m^3. Airborne chromium trioxide concentrations for MIG welders ranged from < 1 to 80 µg/m^3 with a median of 10 µg/m^3. More than half (54%) of measured erythrocyte Cr levels were below the limit of detection (LOD) of 0.6 µg/l. Erythrocyte Cr concentration was recommended for its specificity but limited by its low sensitivity. Chromium was detected in the plasma of all welders, ranging from 2.2 to 68.5 µg/l; approximately 2 to 50 times higher than the level in non-exposed people. Plasma Cr concentration was recommended as a sensitive parameter, limited by its lack of specificity. Erythrocyte, plasma, and urine chromium levels were highly correlated with each other ($P < 0.0001$).

3.3.2 Biological Markers of Effect

3.3.2.1 Renal biomarkers

The concentration levels of certain proteins and enzymes in the urine of workers may indicate early effects of Cr(VI) exposure. Liu et al. [1998] measured urinary N-acetyl-ß-glucosaminidase (NAG), ß$_2$-microglobulin (ß$_2$M), total protein, and microalbumin levels in 34 hard-chrome plating workers, 98 nickel-chrome electroplating workers, and 46 aluminum anode-oxidation workers who had no metal exposure and served as the reference group. Hard-chrome platers were exposed to the highest airborne chromium concentrations (geometric mean 4.20 µg Cr/m^3 TWA) and had the highest urinary NAG concentrations (geometric mean of 4.9 IU/g creatinine). NAG levels were significantly higher among hard-chrome plating workers, while the other biological markers measured were not. NAG levels were significantly associated with age ($P < 0.05$) and gender ($P < 0.01$) and not associated with employment duration.

3.3.2.2 Genotoxic biomarkers

Genotoxic biomarkers may indicate exposure to mutagenic carcinogens. More information about the genotoxic effects of Cr(VI) compounds is presented in Chapter 5, Section 5.2.

Gao et al. [1992] detected DNA strand breaks in human lymphocytes in vitro 3 hours after sodium dichromate incubation, and in male Wistar rat lymphocytes 24 hours after intratracheal instillation. The DNA damage resulting from Cr(VI) exposure is dependent on the reactive intermediates generated [Aiyar et al. 1991].

Gao et al. [1994] investigated DNA damage in the lymphocytes of workers exposed to Cr(VI). No significant increases in DNA strand breaks or 8-OHdG levels were found in the lymphocytes of exposed workers in comparison with the lyphocytes of controls. The exposure level for the exposed group was reported to be approximately 0.01 mg Cr(VI)/m^3.

Gambelunghe et al. [2003] evaluated DNA strand breaks and apoptosis in the peripheral lymphocytes of chrome-plating workers. Previous air monitoring at this plant indicated total chromium levels from 0.4 to 4.5 µg/m^3. Workers exposed to Cr(VI) had higher levels of chromium in their urine, erythrocytes, and lymphocytes than unexposed controls. The comet assay demonstrated an increase in DNA strand breaks in workers exposed to Cr(VI). The percentage of apoptotic nuclei did not differ between exposed workers and controls. Urinary chromium concentrations correlated with erythrocyte chromium concentrations while lymphocyte chromium concentrations correlated with comet tail moment, an indicator of DNA damage.

Kuo et al. [2003] reported positive correlations between urinary 8-OHdG concentrations and both urinary Cr concentration ($P < 0.01$) and airborne Cr concentration ($P < 0.1$) in a study of 50 electroplating workers.

3.3.2.3 Other biomarkers of effect

Li et al. [2001] reported that sperm count and sperm motility were significantly lower ($P < 0.05$) in the semen of workers exposed to Cr(VI) in comparison with the semen of unexposed control workers. The seminal volume and liquefaction time of the semen from the two groups was not significantly different. Workers exposed to Cr(VI) had significantly ($P < 0.05$) increased serum follicle stimulating hormone levels compared with controls; LH and Cr levels were not significantly different between groups. The seminal fluid of exposed workers contained significantly ($P < 0.05$) lower levels of lactate dehydrogenase (LDH), lactate dehydrogenase C4 isoenzyme (LDH-x), and zinc; Cr levels were not different.

4 | Human Health Effects

Most of the health effects associated with occupational hexavalent chromium (Cr[VI]) exposure are well known and have been widely reviewed (see Section 4.1.1, Lung Cancer). The following discussion focuses on quantitative exposure-response studies of those effects and new information not previously reviewed by NIOSH [1975, 1980]. Comprehensive reviews of welding studies are available from ATSDR [2012]; IARC [1990]; and OSHA [71 Fed. Reg. 10099 (2006)]. Analyses of epidemiologic studies with the most robust data for quantitative risk assessment are described in Chapter 6, "Assessment of Risk."

4.1 Cancer

4.1.1 Lung Cancer

Hexavalent chromium is a well-established occupational carcinogen associated with lung cancer and nasal and sinus cancer. In 1989, the International Agency for Research on Cancer (IARC) critically evaluated the published epidemiologic studies of chromium compounds including Cr(VI) and concluded that "there is sufficient evidence in humans for the carcinogenicity of chromium[VI] compounds as encountered in the chromate production, chromate pigment production and chromium plating industries" (i.e., IARC category "Group 1" carcinogen) [IARC 1990]. The IARC-reviewed studies of workers in those industries and the ferrochromium industry are presented in Tables 4–1 through 4–4. (Cr[VI] compounds were reaffirmed as an IARC Group 1 carcinogen [lung] in 2009 [Straif et al. 2009; IARC 2012].) Additional details and reviews of those studies are available in the IARC monograph and elsewhere [IARC 1990; NIOSH 1975a, 1980; WHO 1988; ATSDR 2012; EPA 1998; Dutch Expert Committee on Occupational Standards (DECOS) 1998; Government of Canada et al. 1994; Hughes et al. 1994; Cross et al. 1997; Cohen et al. 1993; Lees 1991; Langård 1983, 1990, 1993; Hayes 1980, 1988, 1997; Gibb et al. 1986; Committee on Biologic Effects of Atmospheric Pollutants 1974]. Although these studies established an association between occupational exposure to chromium and lung cancer, the specific form of chromium responsible for the excess risk of cancer was usually not identified, nor were the effects of tobacco smoking always taken into account. However, the observed excesses of respiratory cancer (i.e., 2-fold to more than 50-fold in chromium production workers) were likely too high to be solely due to smoking.

4.1.1.1 Epidemiologic exposure-response analyses of lung cancer

Sections 4.1.1.1.1 through 4.1.1.1.6 include several epidemiologic studies published after the IARC [1990] review that investigated exposure-response relationships for Cr(VI) and lung cancer using cumulative quantitative Cr(VI) exposure data. Exposure-response models based on cumulative exposure data can predict disease risk for a particular Cr(VI) exposure over a period of time. Epidemiologic studies that provided evidence of an exposure-response relationship based on other kinds of exposure data (e.g., duration of exposure) have been reviewed by the authors cited above and others [CRIOS 2003; K.S. Crump Division 1995]. Reanalyses of data from published epidemiologic studies (i.e., quantitative risk assessments) are described in Chapter 6, "Assessment of Risk," and Chapter 7, "Recommendations for an Exposure Limit."

4.1.1.1.1 U.S. chromate production workers, North Carolina (Pastides et al. [1994a])

A retrospective cohort study of 398 current and former workers employed for at least 1 year from 1971 through 1989 was conducted in a large chromate production facility in Castle Hayne, North Carolina. The plant opened in 1971 and was designed to reduce the high level of chromium exposure found at the company's former production facilities in Ohio and New Jersey. The study was performed to determine if there was early evidence for an increased risk of cancer incidence or mortality and to determine whether any increase was related to the level or duration of exposure to Cr(VI). More than 5,000 personal breathing zone (PBZ) samples collected from 1974 through 1989 were available from company records for 352 of the 398 employees. Concentrations of Cr(VI) ranged from below the limit of detection (LOD) to 289 µg/m³ (8-hour TWA), with > 99% of the samples less than 50 µg/m³. Area samples were used to estimate personal monitoring concentrations for 1971–1972. (Further description of the exposure data is available in Pastides et al. [1994b].) Forty-two of the 45 workers with previous occupational exposure to chromium had transferred from the older Painesville, Ohio plant to Castle Hayne. Estimated airborne chromium concentrations at the Ohio plant ranged from 0.05 mg/m³ to 1.45 mg/m³ of total chromium for production workers to a maximum of 5.67 mg/m³ for maintenance workers (mean not reported).

Mortality of the 311 white male Castle Hayne workers from all causes of death (n = 16), cancer (all sites) (n = 6), or lung cancer (n = 2) did not differ significantly from the mortality experience of eight surrounding North Carolina counties or the United States white male population. Internal comparisons were used to address an apparent "healthy worker" effect in the cohort. Workers with "high" cumulative Cr(VI) exposure (i.e., ≥ 10 "µg-years" of Cr[VI]) were compared with workers who had "low" exposure (i.e., < 10 "µg-years" Cr[VI]). No significant differences in cancer risk were found between the two groups after considering the effects of age, previous chromium exposure, and smoking. There was a significantly increased risk of mortality and cancer, including lung cancer, among a subgroup of employees (11% of the cohort) that transferred from older facilities (odds ratio [OR] for mortality = 1.27 for each 3 years of previous exposure; 90% CI = 1.07–1.51; OR for cancer = 1.22 for each 3 years of previous exposure; 90% CI = 1.03–1.45, controlling for age, years of previous exposure, and smoking status and including malignances among living and deceased subjects). (The authors reported 90% confidence intervals, rather than 95%. Regression analyses that excluded transferred employees were not reported.) The results of this study are limited by a small number of deaths and cases and a short follow-up period, and the authors stated "only a large and early-acting cancer risk would have been identifiable" [Pastides et al. 1994a]. The average total years between first employment in any chromate production facility and death was 15.2 years; the maximum was 35.3 years [Pastides et al. 1994a].

4.1.1.1.2 U.S. chromate production workers, Maryland (Hayes et al. [1979]; Gibb et al. [2000b])

Gibb et al. [2000b] conducted a retrospective analysis of lung cancer mortality in a cohort of

Maryland chromate production workers first studied by Hayes et al. [1979]. The cohort studied by Hayes et al. [1979] consisted of 2,101 male salaried and hourly workers (restricted to 1,803 hourly workers) employed for at least 90 days between January 1, 1945, and December 31, 1974, who had worked in new and/or production sites (Table 4–1). Gibb et al. [2000b] identified a study cohort of 2,357 male workers first employed between 1950 and 1974. Workers who started employment before August 1, 1950, were excluded because a new plant was completed on that date and extensive exposure information began to be collected. Workers starting after that date, but with short-term employment (i.e., < 90 days) were included in the study group to increase the size of the low exposure group. The Hayes et al. [1979] study identified deaths through July 1977. Gibb et al. [2000b] extended the follow-up period until the end of 1992, and included a detailed retrospective assessment of Cr(VI) exposure and information about most workers' smoking habits (see Chapter 6, "Assessment of Risk," for further description of the exposure and smoking data.) The mean length of employment was 3.3 years for white workers (n = 1,205), 3.7 years for nonwhite workers (n = 848), 0.6 years for workers of unknown race (n = 304), and 3.1 years for the total cohort (n = 2,357). The mean follow-up time ranged from 26 years to 32 years; there were 70,736 person-years of observation. The mean cumulative exposures to Cr(VI) were 0.18 mg/m^3-years for nonwhite employees (n = 848) and 0.13 mg/m^3-years for white employees (n = 1,205). The mean exposure concentration was 43 µg/m^3 [Park and Stayner 2006; NIOSH 2005b].

Lung cancer mortality ratios increased with increasing cumulative exposure (i.e., mg CrO$_3$/m^3-years)—from 0.96 in the lowest quartile to 1.57 (95% CI 1.07–2.20; 5-year exposure lag) and 2.24 (95% CI 1.60–3.03; 5-year exposure lag) in the two highest quartiles. The number of expected lung cancer deaths was based on age-, race-, and calendar year-specific rates for Maryland. Proportional hazards models that controlled for the effects of smoking predicted increasing lung cancer risk with increasing Cr(VI) cumulative exposure (relative risks: 1.83 for second exposure quartile, 2.48 for third exposure quartile, and 3.32 for fourth exposure quartile, compared with first quartile of cumulative exposure; confidence intervals not reported; 5-year exposure lag) [Gibb et al. 2000b]. For further exploration of time and exposure variables and lung cancer mortality see Gibb et al. [2011].

In an analysis by industry consultants of simulated cohort data, lung cancer mortality ratios remained statistically significant for white workers and the total cohort regardless of whether city, county, or state reference populations were used [Exponent 2002]. The simulated data were based on descriptive statistics for the entire cohort provided in Gibb et al. [2000b], mainly Table 2.

4.1.1.1.3 U.S. chromate production workers, Ohio (Luippold et al. [2003])

Luippold et al. [2003] conducted a retrospective cohort study of lung cancer mortality in 482 chromate production workers (four female workers) employed ≥ 1 year from 1940 through 1972 in a Painesville, Ohio plant studied earlier by Mancuso [1975, 1997]. The current study identified a more recent cohort that did not overlap with the Mancuso cohorts. These workers had not been employed in any of the company's other facilities that used or produced Cr(VI). However, workers who later worked at the North Carolina plant that had available quantitative estimates of Cr(VI) were included in this study. The number included was not reported in Luippold et al. [2003]; Proctor et al. [2004] stated that 17 workers who transferred to the North

Carolina plant had their exposure profiles incorporated. Their mortality was followed from 1941 through 1997 and compared with United States and Ohio rates. Nearly half (i.e., 45%) of the cohort worked in exposed jobs for 1 to 4 years; 16% worked in them > 20 years. Follow-up length averaged 30 years, ranging from 1 to 58 years. However, of the workers who died from lung cancer (n = 51), 43% worked 20 or more years and 82% began plant employment before 1955. Their follow-up length averaged 31.6 years, ranging from 7 to 52 years and totaling 14,048 person-years. More than 800 area samples of airborne Cr(VI) from 21 industrial hygiene surveys were available for formation of a job-exposure matrix. The surveys were conducted in 1943, 1945, 1948, and every year from 1955 through 1971. Samples were collected in impingers and analyzed colorimetrically for Cr(VI). Concentrations tended to decrease over time. The average airborne concentration of Cr(VI) in the indoor operating areas of the plant in the 1940s was 0.72 mg/m^3, 0.27 mg/m^3 from 1957 through 1964, and 0.039 mg/m^3 from 1965 through 1972 [Proctor et al. 2003]. Further details about the exposure data are in Proctor et al. [2003]. For the lung cancer deaths, mean cumulative Cr(VI) exposure was 1.58 mg/m^3-years (range: 0.003–23 mg/m^3-years) for the cohort and 3.28 mg/m^3-years (range: 0.06–23 mg/m^3-years). The effects of smoking could not be assessed because of insufficient data.

Cumulative Cr(VI) exposure was divided into five categories to allow for nearly equal numbers of expected deaths from cancer of the trachea, bronchus, or lung in each category: 0.00–0.19, 0.20–0.48, 0.49–1.04, 1.05–2.69, and 2.70–23.0 mg/m^3-years [Luippold et al. 2003]. Person-years in each category ranged from 2,369 to 3,220, and the number of deaths from trachea, bronchus, or lung cancer ranged from 3 in the lowest exposure category to 20 in the highest (n = 51). The standardized mortality ratios (SMRs) were statistically significant in the two highest cumulative exposure categories (3.65 [95% CI 2.08–5.92] and 4.63 [2.83–7.16], respectively). SMRs were also significantly increased for year of hire before 1960, ≥ 20 years of employment, and ≥ 20 years since first exposure. The tests for trend across increasing categories of cumulative Cr(VI) exposure, year of hire, and duration of employment were statistically significant ($P \leq 0.005$). A test for departure of the data from linearity was not statistically significant (χ^2 goodness of fit of linear model; $P = 0.23$). Van Wijngaarden et al. [2004] reported further examination and discussion of cumulative Cr(VI) exposure and lung cancer mortality in this study and Gibb et al. [2000b].

4.1.1.1.4 U.S. chromate production workers (Luippold et al. [2005])

Luippold et al. [2005] conducted a retrospective cohort mortality study of 617 male and female chromate production workers employed at least 1 year at one of two U.S. plants: 430 workers from the North Carolina plant studied by Pastides et al. [1994a] (i.e., "Plant 1") and 187 workers hired after the 1980 implementation of exposure-reducing process changes at "Plant 2". The study's primary goal was to investigate possible cancer mortality risks associated with Cr(VI) exposure after production process changes and enhanced industrial hygiene controls (i.e., the "postchange environment"). Employees who had worked less than 1 year in a postchange plant or in a facility using a high-lime process were excluded from the cohort. Personal air-monitoring measurements available from 1974 to 1988 for Plant 1 and from 1981 through 1998 for Plant 2 indicated that, for most years, overall geometric mean Cr(VI) concentrations for both plants were less than 1.5 µg/m^3 and area-specific average personal air-sampling values were generally less than 10 µg/m^3.

Cohort mortality was followed through 1998. The mean time since first exposure was 20.1 years for Plant 1 workers and 10.1 years for Plant 2. Only 27 cohort members (4%) were deceased, and stratified analyses with individual exposure estimates and available smoking history data could not be conducted because of the small number of deaths. Mortality from all causes was lower than the expected number of deaths based on state-specific referent rates, suggesting a strong healthy worker effect (SMR 0.59; 95% CI 0.39–0.85; 27 deaths). Lung cancer mortality was also lower than expected compared with state reference rates (SMR 0.84; 95% CI 0.17–2.44; 3 deaths). However, the study results are limited by a small number of deaths and short follow-up period. The authors stated that the "absence of an elevated lung cancer risk may be a favorable reflection of the postchange environment," but longer follow-up allowing an appropriate latency period for the entire cohort is needed to confirm this preliminary conclusion [Luippold et al. 2005].

4.1.1.1.5 Chromate production workers, Germany (Birk et al. [2006])

Birk et al. [2006] conducted a retrospective cohort study of lung cancer mortality using Cr levels in urine as a biomarker of occupational exposure to Cr(VI). Cohort members were males employed in two German chromate production plants after each plant converted to a no-lime production process, a process believed to result in dusts containing less Cr(VI) [Birk et al. 2006]. The average duration of Cr(VI) exposure was 9–11 years and mean time since first exposure was 16–19 years, depending on the plant (i.e., Plant A or Plant B). Smoking status from medical examinations/medical records was available for > 90% of the cohort as were > 12,000 urinary chromium results collected during routine employee medical examinations of workers from both plants.

Mortality was followed through 1998; 130 deaths (22 deaths from cancer of the trachea, bronchus, or lung) were identified among 901 workers employed at least 1 year in the plant with no history of work in a plant before conversion to the no-lime process. The number of person-years was 14,684. Although mortality from all causes was significantly less than the expected number compared with mortality rates for Germany, the number of deaths from cancer of the trachea, bronchus, or lung was greater than expected (SMR = 1.48; 22 deaths observed; 14.83 expected; 95% CI 0.93–2.25). When regional mortality rates were used (i.e., North Rhine-Westphalia), the SMRs were somewhat lower (SMR for all respiratory cancers including trachea, bronchus, and lung = 1.22; 95% CI 0.76–1.85).

Geometric mean values of Cr in urine varied by work location, plant, and time period, and tended to decrease over the years of plant operation (both plants are now closed). Results of statistical analysis found lung cancer mortality SMRs > 2.00 in the highest cumulative Cr-in-urine exposure category, for no exposure lag, 10-year lag, and 20-year lag (e.g., a statistically significant highest SMR was reported in the highest exposure category of ≥ 200 µg/L-years Cr in urine: SMR 2.09; 12 lung cancer deaths observed; 95% CI 1.08–3.65; regional rates; no exposure lag). However, few study subjects accrued high cumulative exposures of 20 years or more before the end of the study. Cumulative urinary Cr concentrations of ≥ 200 µg/L-years compared with concentrations < 200 µg/L-years were associated with a significantly increased risk of lung cancer mortality (OR = 6.9; 95% CI 2.6–18.2), and the risk was unchanged after controlling for smoking [Birk et al. 2006].

The use of urinary Cr measurements as a marker for Cr(VI) exposure has limitations, primarily that it may reflect exposure to Cr(VI), Cr(III), or both. In addition, urinary Cr levels may reflect beer consumption or smoking; however, the study authors stated that "... workplace exposures to hexavalent chromium are expected to have a much greater impact on overall urinary chromium levels than normal variability across individuals due to dietary and metabolic differences" [Birk et al. 2006].

4.1.1.1.6 European welders (Simonato et al. [1991])

IARC researchers conducted a large study of lung cancer in 11,092 male welders (164,077 person-years) from 135 companies in nine European countries. Stainless steel welders are exposed to welding fumes that can contain Cr(VI) and other carcinogens such as nickel. Mortality and incidence were analyzed by cause, time since first exposure, duration of employment, and estimated cumulative exposure to total fumes, chromium (Cr), Cr(VI), and nickel (Ni). The observation period and criteria for inclusion of welders varied from country to country. Data about subjects' smoking habits were not available for the entire cohort, so no adjustment could be made. While mortality from all causes of death was significantly lower than national rates, the number of deaths from lung cancer (116 observed; 86.81 expected; SMR 1.34 [95% CI 1.10–1.60]), and malignant neoplasms of the bladder (15 observed; 7.86 expected; SMR 1.91 [95% CI 1.07–3.15]) were significantly higher. Lung cancer SMRs tended to increase with years since first exposure for stainless steel welders and mild steel welders; the trend was statistically significant for the stainless steel welders ($P < 0.05$). The SMRs for subgroups of stainless steel welders with at least 5 years of employment and 20 years since first exposure and high cumulative exposure to either Cr(VI) or Ni (i.e., ≥ 0.5 mg-years/m^3) were not significantly higher than SMRs for the low cumulative exposure subgroup (i.e., < 0.5 mg-years/m^3) [Simonato et al. 1991].

IARC classifies welding fumes and gases as Group 2B carcinogens—limited evidence of carcinogenicity in humans [IARC 1990]. During a 2009 review, IARC found sufficient evidence for ocular melanoma in welders [El Ghissassi et al. 2009]. NIOSH recommends that "exposures to all welding emissions be reduced to the lowest feasible concentrations using state-of-the-art engineering controls and work practices" [NIOSH 1988a].

4.1.2 Nasal and Sinus Cancer

Cases or deaths from sinonasal cancers were reported in five IARC-reviewed studies of chromium production workers in the United States, United Kingdom, and Japan, chromate pigment production workers in Norway, and chromium platers in the United Kingdom (see Tables 4–1 through 4–3). IARC concluded that the findings represented a "pattern of excess risk" for these rare cancers [IARC 1990] and in 2009 concluded there is limited evidence for human cancers of the nasal cavity and paranasal sinuses from exposure to Cr(VI) compounds [Straif et al. 2009; IARC 2012].

Subsequent mortality studies of chromium or chromate production workers employed in New Jersey from 1937 through 1971 and in the United Kingdom from 1950 through 1976 reported significant excesses of deaths from nasal and sinus cancer (proportionate cancer mortality ratio (PCMR) = 5.18 for white males, $P < 0.05$, six deaths observed and no deaths observed in black males [Rosenman and Stanbury 1996]; SMR adjusted for social class and area = 1,538, $P < 0.05$, four deaths observed [Davies et al. 1991]). Cr(VI) exposure concentrations were not reported. However, an earlier survey of three chromate production facilities in the U.K. found that average air concentrations of Cr(VI) in various phases of the process ranged from 0.002 to 0.88 mg/m^3 [Buckell and Harvey 1951; ATSDR 2012].

Four cases of carcinoma of the nasal region were described in male workers with 19 to 32 years of employment in a Japanese chromate factory [Satoh et al. 1994]. No exposure concentrations were reported.

Although increased or statistically significant numbers of cases of nasal or sinonasal cancer have been reported in case-control or incidence studies of leather workers (e.g., boot and shoe production) or leather tanning workers in Sweden and Italy [Comba et al. 1992; Battista et al. 1995; Mikoczy and Hagmar 2005], a U.S. mortality study did not find an excess number of deaths from cancer of the nasal cavity [Stern et al. 2003]. The studies did not report quantitative exposure concentrations of Cr(VI), and a causative agent could not be determined. Leather tanning workers may be exposed to several other potential occupational carcinogens, including formaldehyde.

4.1.3 Nonrespiratory Cancers

Statistically significant excesses of cancer of the oral region, liver, esophagus, and all cancer sites combined were reported in a few studies reviewed by IARC (Tables 4–1 through 4–4). IARC [1990] concluded that "for cancers other than of the lung and sinonasal cavity, no consistent pattern of cancer risk has been shown among workers exposed to chromium compounds." More recent reviews by other groups also did not find a consistent pattern of nonrespiratory cancer risk in workers exposed to inhaled Cr(VI) [ATSDR 2012; Proctor et al. 2002; Chromate Toxicity Review 2001; EPA 1998; Government of Canada 1994; Cross et al. 1997; CRIOS 2003; Criteria Group for Occupational Standards 2000]. IARC [2012] concluded that "there is little evidence that exposure to chromium (VI) causes stomach or other cancers."

4.1.4 Cancer Meta-Analyses

Meta-analysis and other systematic literature review methods are useful tools for summarizing exposure risk estimates from multiple studies. Meta-analyses or summary reviews of epidemiologic studies have been conducted to investigate cancer risk in chromium-exposed workers.

Steenland et al. [1996] reported overall relative risks for specific occupational lung carcinogens, including chromium. Ten epidemiologic studies were selected by the authors as the largest and best-designed studies of chromium production workers, chromate pigment production workers, and chromium platers (i.e., Enterline 1974; Hayes et al. 1979; Alderson et al. 1981; Satoh et al. 1981; Korallus et al. 1982; Frentzel-Beyme 1983; Davies 1984; Sorahan et al. 1987; Hayes et al. 1989; Takahashi and Okubo 1990).

The summary relative risk for the 10 studies was 2.78 (95% CI 2.47–3.52; random effects model), which was the second-highest relative risk among eight carcinogens summarized.

Cole and Rodu [2005] conducted meta-analyses of epidemiologic studies published in 1950 or later to test for an association of chromium exposure with all causes of death, and death from malignant diseases (i.e., all cancers combined, lung cancer, stomach cancer, cancer of the central nervous system [CNS], kidney cancer, prostate gland cancer, leukemia, Hodgkin's disease, and other lymphohematopoietic cancers). Available papers (n = 114) were evaluated independently by both authors on eight criteria that addressed study quality. In addition, papers with data on lung cancer were assessed for control of cigarette smoking, and papers with data on stomach cancer were assessed for economic status. Lung or stomach cancer papers that were negative or "essentially negative" regarding chrome exposure were included with papers that controlled for smoking or economic status. Forty-nine epidemiologic studies, based on 84 papers published since 1950, were used in the meta-analyses.

The number of studies in each meta-analysis ranged from 9 for Hodgkin's disease to 47 for lung cancer. Most studies investigated occupational exposure to chromium. Association was measured by an author-defined "SMR," which included odds ratios, proportionate mortality ratios and, most often, standardized mortality ratios. Confidence intervals (i.e., 95%) were calculated by the authors. Mortality risks were not significantly increased for most causes of death (i.e., all causes, prostate gland cancer, kidney cancer, CNS cancer, leukemia, Hodgkin's disease, or other lymphohematopoietic cancers). However, SMRs were significantly increased in all lung cancer meta-analyses (smoking controlled: 26 studies; 1,325 deaths; SMR = 118; 95% CI 112–125) (smoking not controlled: 21 studies; 1,129 deaths; SMR = 181; 95% CI 171–192) (lung cancer—all: 47 studies; 2,454 deaths; SMR = 141; 95% CI 135–147). Stomach cancer mortality risk was significantly increased only in meta-analyses of studies that did not control for effects of economic status (economic status not controlled: 18 studies; 324 deaths; SMR = 137; 95% 123–153). The authors stated that statistically significant SMRs for "all cancer" mortality were mainly due to lung cancer (all cancer: 40 studies; 6,011 deaths; SMR = 112; 95% CI 109–115). Many of the studies contributing to the meta-analyses did not address bias from the healthy worker effect, and thus the results are likely underestimates of the cancer mortality risks. Other limitations of these meta-analyses include lack of (1) exposure characterization of populations such as the route of exposure (i.e., airborne versus ingestion) and (2) detail of criteria used to exclude studies based on "no or little chrome exposure" or "no usable data."

Paddle [1997] conducted a meta-analysis of four studies of chromate production workers in plants in the United States (Hayes et al. [1979]; Pastides et al. [1994a]), United Kingdom (i.e., Davies et al. [1991]), and Germany (i.e., Korallus et al. [1993]) that had undergone modifications to reduce chromium exposure. Most of the modifications occurred around 1960. This meta-analysis of lung cancer "post-modification" did not find a statistically significant excess of lung cancer (30 deaths observed; 27.2 expected; risk measure and confidence interval not reported). The author surmised that none of the individual studies in the meta-analysis or the meta-analysis itself had sufficient statistical power to detect a lung cancer risk of moderate size because of the need to exclude employees who worked before plant modifications and the need to incorporate a latency period, thus leading to very small observed and expected numbers. Meta-analyses of gastrointestinal cancer, laryngeal cancer, or any other nonlung cancer were considered inappropriate by the author because of reporting bias and inconsistent descriptions of the cancer sites [Paddle 1997].

Sjögren et al. [1994] authored a brief report of their meta-analysis of five lung cancer studies of Canadian and European welders exposed to stainless steel welding fumes. The meta-analysis found an estimated relative risk of 1.94 (95% CI 1.28–2.93) and accounted for the effects of smoking and asbestos exposure [Sjögren et al. 1994]. (Details of each study's exposure assessment and concentrations were not included.)

4.1.5 Summary of Cancer and Cr(VI) Exposure

Occupational exposure to Cr(VI) has long been associated with nasal and sinus cancer and cancers of the lung, trachea, and bronchus. No consistent pattern of nonrespiratory cancer risk has been identified.

Few studies of Cr(VI) workers had sufficient data to determine the quantitative relationship between cumulative Cr(VI) exposure and lung cancer risk while controlling for the effects of other lung carcinogens, such as tobacco smoke. One such study found a significant relationship between cumulative Cr(VI) exposure (measured as CrO_3) and lung cancer mortality [Gibb et al. 2000b]; the data were reanalyzed by NIOSH to further investigate the exposure-response relationship (see Chapter 6, "Assessment of Risk").

The three meta-analyses and summary reviews of epidemiologic studies with sufficient statistical power found significantly increased lung cancer risks with chromium exposure.

4.2 Nonmalignant Effects

Cr(VI) exposure is associated with contact dermatitis, skin ulcers, irritation and ulceration of the nasal mucosa, and perforation of the nasal septum [NIOSH 1975a]. Reports of kidney damage, liver damage, pulmonary congestion and edema, epigastric pain, erosion and discoloration of teeth, and perforated ear drums were found in the literature, and NIOSH concluded that "sufficient contact with any chromium(VI) material could cause these effects" [NIOSH 1975a]. Later studies that provided quantitative Cr(VI) information about the occurrence of those effects is discussed here. (Studies of nonmalignant health effects and total chromium concentrations [i.e., non-speciated] are included in reviews by the Criteria Group for Occupational Standards [2000] and ATSDR [2012].)

4.2.1 Respiratory Effects

The ATSDR [2012] review found many reports and studies published from 1939 to 1991 of workers exposed to Cr(VI) compounds for intermediate (i.e., 15–364 days) to chronic durations that noted these respiratory effects: epistaxis, chronic rhinorrhea, nasal itching and soreness, nasal mucosal atrophy, perforations and ulcerations of the nasal septum, bronchitis, pneumoconiosis, decreased pulmonary function, and pneumonia.

Five recent epidemiologic studies of three cohorts analyzed quantitative information about occupational exposures to Cr(VI) and respiratory effects. The three worksite surveys described below provide information about workplace Cr(VI) concentrations and health effects at a particular point in time only and do not include statistical analysis of the quantitative relationship between specific work exposures and reported health symptoms; thus they contribute little to evaluation of the exposure-response association. (Studies and surveys previously reviewed by NIOSH [1975, 1980] are not included.)

4.2.1.1 Work site surveys

A NIOSH HHE of 11 male employees in an Ohio electroplating facility reported that most men had worked in the "hard-chrome" area for the majority of their employment (average duration: 7.5 years; range: 3–16 years). Four of the 11 workers had a perforated nasal septum. Nine of the 11 men had hand scars resulting from past chrome ulcerations. Other effects found during the investigation included nose bleeds, "runny nose," and nasal ulcerations. A total of 17 air samples were collected for Cr(VI) with a vacuum pump in 2 days during periods of 2–4 hours (14 personal; 3 area). The mean Cr(VI) concentration was 0.004 mg/m^3 (range: < 0.001 mg/m^3–0.02 mg/m^3) [NIOSH 1975c]. This survey focused on chromic acid exposure; other potential exposures were not noted in the report. Possible limitations of this study include (1) lack of a comparison or unexposed "control" group, (2) inclusion of only current workers, and (3) a small and possibly unrepresentative study group. Other NIOSH HHEs that noted nasal sores or other respiratory effects in workers exposed to chromium had similar limitations and are not discussed here. In addition, some surveys were conducted in workplaces with air concentrations of chromium and other metals, dusts, and chemicals (e.g., nickel, copper, zinc, particulates, ammonia [NIOSH 1985a,b], sulfur dioxide, welding fume, aluminum, carbon monoxide, nitrogen dioxide [Burkhart and Knutti 1994]) that could have contributed to observed and reported effects.

An HHE at a small chrome plating shop with six workers (including four platers) found among the workers no nasal ulcerations, nasal septal perforations, or lesions on the hands. However, information was obtained by interview, observation, and questionnaire, and no medical examinations were performed. Four PBZ samples with durations of 491 to 505 minutes were analyzed and found to contain low air concentrations of Cr(VI) (0.003–0.006 mg/m^3) and total chromium (0.009–0.011 mg/m^3). The HHE was requested because of reported overexposure to chemicals used in chrome plating, poor ventilation, and cardiovascular disorders among employees. NIOSH determined that (1) overexposures to plating chemicals did not exist, (2) local exhaust systems were operating "below recommended levels," and (3) no occupational factors contributing to heart disease were identified. Recommendations were made for ventilation, housekeeping, and personal protective equipment (PPE) [Ahrenholz and Anderson 1981].

Eleven cases of nasal septum perforation were found in 2,869 shipyard welders in Korea [Lee et al. 2002]. The affected workers had no history of trauma, surgery, diseases, or medication use that could account for the perforations. Blood and urine chrome concentrations of the affected workers were below the LOD. The affected workers ranged in age from 37 to 51 years and had welded 12–25 years. Personal air samples for Cr(VI) were collected from 31 workers in a stainless steel welding shop (shop F) and the five work locations (i.e., CO_2 welding shops A–E), where the 11 workers who had septum perforation were last employed. ("Most" of the workers had not recently worked in shop F.) Mean, maximum, and minimum Cr(VI) concentrations, and number of affected workers were reported for each shop (shops A, B, D, and E had two affected workers; shop C had three). The number of unaffected workers (non-cases) per shop was not reported. The mean concentrations of Cr(VI) in the welding fume ranged from 0.0012 mg/m^3 (shop B) to 0.22 mg/m^3 (8-hour TWA) in shop F. The highest maximum (0.34 mg/m^3) and minimum (0.044 mg/m^3) Cr(VI) concentrations were also measured in shop F. The mean Cr(VI) concentrations in shops A, C, D and E ranged from 0.0014 (shop C) to 0.0028 mg/m^3 (shop E) (maximums for A–E: 0.0013 mg/m^3–0.0050 mg/m^3). Annual industrial hygiene surveys for air concentrations of metals conducted from 1991 through 2000 found that mean total "chrome" (i.e., Cr) concentrations ranged from 0.002 to 0.025 mg/m^3, and the maximum concentrations were 0.010–0.509 mg/m^3. The authors judged that pre-1990 concentrations were higher. The authors could not obtain annual total Cr or Cr(VI) concentrations for the stainless steel welding workplace. Use of a comparison group was not reported. The authors assumed that the nasal septal perforations were caused by "long-term exposure to the low levels of hexavalent chromium during welding" [Lee et al. 2002] (mean concentrations exceeded the existing and revised Cr(VI) RELs).

4.2.1.2 Epidemiologic studies

4.2.1.2.1 Lindberg and Hedenstierna [1983]

A cross-sectional study of respiratory symptoms, changes in nasal mucosa, and lung function was conducted in chrome-plating workers in Swedish factories (n = 43: 16 male nonsmokers; 21 male smokers; 3 female nonsmokers; 3 female smokers) [Lindberg and Hedenstierna 1983]. Five chrome baths in three factories were studied for a total of 19 work days. Office employees (n = 19: 13 males; 14 nonsmokers) and auto mechanics (n = 119 males; 52 nonsmokers) were used as comparison groups for nose and throat effects, and lung function, respectively. For analysis of subjective symptoms and nasal conditions, the 43 exposed workers were divided into two groups: "low" exposure (8-hour mean ≤ 1.9 µg/m^3 chromic acid; 19 workers) and "high" mean exposure (2–20 µg/m^3 chromic acid; 24 workers). Mean daily Cr(VI) exposures ranged from ≤ 1.9 to 20 µg/m^3. Their median duration of employment was 2.5 years (range: 0.2–23.6 years). Exposure concentrations were measured with personal air samplers and stationary equipment placed near the chromic acid baths. A statistically significant difference was found in the low exposure group when compared with controls for the effect of "smeary and crusty septal mucosa" (11/19 workers versus 5/19 controls; $P < 0.05$). There were no perforations or ulcerations in the low exposure group. Frequency of nasal atrophy was significantly greater in the high exposure group compared with the controls (8/24 workers versus 0/19 controls; $P < 0.05$). The high exposure group also had higher frequency of nasal mucosal ulcerations and/or septal perforations (8 workers with ulcerations—2 of those also had perforations; 5 workers with perforations—2 of those also had ulcerations; $P < 0.01$; number of controls not reported). Fourteen workers were temporarily exposed to peak concentrations of 20–46 µg/m^3 when working near the baths; 10 of those workers had nasal mucosal ulcerations with or without perforation, or perforation only. Workers with low exposure had no significant changes in lung function during the survey. Workers in the high exposure group had slight transient decreases in forced vital capacity (FVC), forced expired volume in one second (FEV$_1$) and forced mid-expiratory flow during the work week.

The results of that study were used by ATSDR to determine an inhalation minimum risk level (MRL) of 0.000005 mg/m^3 (0.005 µg/m^3) for intermediate-duration exposure (15–364 days) to Cr(VI) as chromium trioxide mist and other dissolved Cr(VI) aerosols and mists. (An intermediate-duration inhalation MRL of 0.0003 mg Cr(VI)/m^3 for exposure to chromium (VI) particulates was derived from a study in rats, Glaser et al. [1990].)

4.2.1.2.2 Huvinen et al. [1996; 2002a,b]

No increased prevalences of respiratory symptoms, lung function deficits, or signs of pneumoconiosis (i.e., small radiographic opacities) were found in a 1993 cross-sectional study of stainless steel production workers [Huvinen et al. 1996]. The median personal Cr(VI) concentration measured in the steel smelting shop in 1987 was 0.5 µg/m^3 (i.e., 0.0005 mg/m^3). (Duration of sample collection and median Cr(VI) concentrations for other work areas were not reported.) The study group consisted of 221 production workers with at least 8 years of employment in the same department and a control group of 95 workers from the cold rolling mill and other areas where chromium or dust exposure was minimal or non-existent. The chromium-exposed workers were divided into three groups: Cr(VI)-exposed (n = 109), Cr(III)-exposed (n = 76), and chromite-exposed (n = 36). Questionnaires regarding health symptoms were completed by 37 former workers; none of those workers reported leaving the company because of a disease. One person reported having chronic bronchitis, two reported having bronchial asthma, and no former workers reported other pulmonary diseases, allergic rhinitis, or cancer. Controls and workers exposed to Cr(VI) had similar mean durations of employment (exposed: 16.0 years; controls: 14.4 years), smoking habits, and other characteristics. Logistic regression analyses adjusted for effects of confounding factors and found no significant differences between Cr(VI)-exposed workers and controls in reported symptom prevalences, prevalence of impaired lung function (with the exception of impaired peak expiratory flow which was significantly more prevalent in the control group [$P < 0.05$]), or occurrence of small opacities.

A similar cross-sectional study of the same cohort 5 years later yielded similar results [Huvinen et al. 2002a]. The median Cr(VI) personal concentration (duration of sample collection time not reported) measured in the steel smelting shop in 1999 had decreased to 0.0003 mg/m³ (maximum: 0.0007 mg/m³), which the authors attributed to technological improvements in production processes. (Exposure concentrations reported in the text and tables differed; table values are reported here.) Cr(VI)-exposed workers (n = 104; mean duration of employment: 21.0 years) and controls (n = 81; mean employment: 19.4 years) did not differ significantly in prevalence of respiratory symptoms or lung function deficits. The profusion of small opacities had progressed in three workers (ILO category ≥ 1/0), including one exposed to Cr(VI). Based on the findings in both studies, the authors concluded that exposure to chromium compounds at the measured concentrations does not produce pulmonary fibrosis. Clinical examinations of 29 workers exposed to Cr(VI) from the steel smelting shop found no nasal tumors, chronic ulcerations, or septal perforations (mean duration of employment: 21.4 years) [Huvinen et al. 2002b].

4.2.1.2.3 Gibb et al. [2000a]

A retrospective study of 2,357 males first employed from 1950 through 1974 at a chromate production plant included a review of clinic and first aid records for physician findings of nasal irritation, ulceration, perforation; and bleeding, skin irritation and ulceration, dermatitis, burns, conjunctivitis, and perforated eardrum [Gibb et al. 2000a]. The mean and median annual airborne Cr(VI) concentrations (measured as CrO_3) for the job title where the clinical finding first occurred and cohort percentages with various clinical findings, from start of employment to occurrence of the first finding, were determined. (See Chapter 6 for further description of the exposure data.) About 40% of the cohort (n = 990) worked fewer than 90 days. These short-term workers were included to increase the low exposure group. Medical records were available for 2,307 men (97.9% of total cohort). The record review found that more than 60% of the cohort had irritated nasal septum (68.1%) or ulcerated nasal septum (62.9%). Median Cr(VI) exposure (measured as CrO_3) at the time of first diagnosis of these findings and all others (i.e., perforated nasal septum, bleeding nasal septum, irritated skin, ulcerated skin, dermatitis, burn, conjunctivitis, and perforated eardrum) was 0.020–0.028 mg/m³ (20–28 µg/m³). The median time from date first employed to date of first diagnosis was less than 1 month for three conditions: irritated nasal septum (20 days), ulcerated nasal septum (22 days), and perforated eardrum (10 days). The mean time from date first employed to date of first diagnosis for each of these conditions was 89 days for irritated nasal septum, 86 days for ulcerated nasal septum, and 235 days for perforated eardrum. The relationship between Cr(VI) exposure and first occurrence of each clinical finding was evaluated with a proportional hazards model. The model predicted that ambient Cr(VI) exposure was significantly associated with occurrence of ulcerated nasal septum ($P = 0.0001$), ulcerated skin ($P = 0.004$), and perforated eardrum ($P = 0.03$). Relative risks per 0.1 mg/m³ increase in CrO_3 were 1.20 for ulcerated nasal septum, 1.11 for ulcerated skin, and 1.35 for "perforated ear." Calendar year of hire was associated with each finding except conjunctivitis and irritated skin; the risk decreased as year of hire became more recent. The authors suggested that the reduction could possibly be because of decreases in ambient Cr(VI) exposure from 1950 to 1985 or changes in plant conditions, such as use of respirators and personal hygiene measures [Gibb et al. 2000a]. The authors also suggested that the proportional hazards model did not find significant associations with all symptoms because the Cr(VI) concentrations were based on annual averages rather than on shorter, more recent average exposures, which may have been a more relevant choice.

4.2.1.3 Summary of respiratory effects studies and surveys

A few workplace surveys measured Cr(VI) air concentrations and conducted medical evaluations of workers. These short-term surveys did not include comparison groups or exposure-response analyses. Two surveys found U.S. electroplaters and Korean welders with nasal perforations or other respiratory effects; the lowest mean Cr(VI) concentrations at the worksites were 0.004 mg/m³ for U.S. electroplaters and 0.0012 mg/m³ for Korean welders [NIOSH 1975c; Lee et al. 2002].

Cross-sectional epidemiologic studies of chrome-plating workers [Lindberg and Hedenstierna 1983] and stainless steel production workers [Huvinen et al. 1996; 2002a,b] found no nasal perforations at average chromic acid concentrations < 2 µg/m³. The platers experienced nasal ulcerations and/or septal perforations and transient reductions in lung function at mean concentrations ranging from 2 µg/m³ to 20 µg/m³. Nasal mucosal ulcerations and/or septal perforations occurred in plating workers exposed to peak concentrations of 20–46 µg/m³.

The best exposure-response information to date is from the only epidemiologic study with sufficient health and exposure data to estimate the risks of ulcerated nasal septum, ulcerated skin, perforated nasal septum, and perforated eardrum over time [i.e., Gibb et al. 2000a]. This retrospective study reviewed medical records of more than 2,000 male workers and analyzed thousands of airborne Cr(VI) measurements collected from 1950 through 1985. More than 60% of the cohort had experienced an irritated nasal septum (68.1%) or ulcerated nasal septum (62.9%) at some time during their employment. The median Cr(VI) exposure (measured as CrO_3) at the time of first diagnosis of these findings and all others (i.e., perforated nasal septum, bleeding nasal septum, irritated skin, ulcerated skin, dermatitis, burn, conjunctivitis, perforated eardrum) was 0.020 mg/m³–0.028 mg/m³ (20 µg/m³–28 µg/m³). Of particular concern is the finding of nasal and ear effects occurring in less than 1 month: the median time from date first employed to date of first diagnosis was less than 1 month for irritated nasal septum (20 days), ulcerated nasal septum (22 days), and perforated eardrum (10 days). A proportional hazards model predicted relative risks of 1.20 for ulcerated nasal septum, 1.11 for ulcerated skin, and 1.35 for "perforated ear" for each 0.1 mg/m³ increase in ambient CrO_3. The authors noted that the chrome platers studied by Lindberg and Hedenstierna [1983] were exposed to chromic acid, which may be more irritative than the chromate chemicals occurring with chromate production [Gibb et al. 2000a].

4.2.1.4 Asthma

Occupational asthma caused by chromium exposure occurs infrequently compared with allergic contact dermatitis [Leroyer et al. 1998]. The exposure concentration below which no cases of occupational asthma would occur, including cases induced by chromium compounds, is not known [Chan-Yeung 1995]. Furthermore, that concentration is likely to be lower than the concentration that initially led to the employee's sensitization [Chan-Yeung 1995]. Case series of asthma have been reported in U.K. electroplaters [Bright et al. 1997], Finnish stainless steel welders [Keskinen et al. 1980], Russian alumina industry workers [Budanova 1980]; Korean metal plating, construction, and cement manufacturing workers, [Park et al. 1994]; and a cross-sectional study of U.K. electroplaters [Burges et al. 1994]. However, there are no quantitative exposure-response assessments of asthma related to Cr(VI) in occupational cohorts, and further research is needed.

4.2.2 Dermatologic Effects

Cr(VI) compounds can cause skin irritation, skin ulcers, skin sensitization, and allergic contact dermatitis. In 1975, NIOSH recommended protective clothing and other measures to prevent occupational exposure [NIOSH 1975a]. Because of those health hazards, potential eye contact, or other nonrespiratory hazards, protective measures and appropriate work practices are recommended "regardless of the airborne concentration of chromium(VI)" [NIOSH 1975a]. Current recommendations for prevention of dermal exposure to Cr(VI) compounds are presented in Chapter 8, "Risk Management."

There are many occupational sources of chromium compounds. Dermatologic effects (i.e., mainly allergic contact dermatitis [ACD]) have been reported from exposure to cement and cement-hardening agents, cleaning, washing, and bleaching materials, textiles and furs, leather and artificial leather tanned with chromium, chrome baths, chromium ore, chrome colors and dyes, pigments in soaps, primer paints, anti-corrosion agents, cutting fluids, machine oils, lubricating oils and greases, glues, resin hardeners, wood preservatives, boiler linings, foundry sand, matches, welding fumes, and other sources [Burrows et al. 1999; Burrows 1983, 1987; Handley and Burrows 1994; Haines and Nieboer 1988; Polak 1983].

No occupational studies have examined the quantitative exposure-response relationship between Cr(VI) exposure and a specific dermatologic effect, such as ACD; thus, an exposure-response relationship has not been clearly established.

Gibb et al. [2000a] evaluated mean Cr(VI) exposure and mean and median time from first employment to the diagnosis of several skin or membrane irritations: irritated skin, ulcerated skin, dermatitis, burn, and conjunctivitis (see Sections 4.2.1.2 and 4.2.1.3). Ulcerated skin and burns were reported in more than 30% of the cohort. The mean Cr(VI) concentration (measured as CrO_3) ranged from 0.049 mg/m^3 to 0.058 mg/m^3 at the time of first diagnosis of those five effects. The mean days on the job until first diagnosis ranged from 373 to 719 days (median 110–221 days).

Other assessments evaluated the occurrence of ACD from contact with Cr(VI) in soil (e.g., Proctor et al. [1998]; Paustenbach et al. [1992]; Bagdon and Hazen [1991]; Stern et al. [1993]; Nethercott et al. [1994, 1995]).

4.2.3 Reproductive Effects

The six available studies of pregnancy occurrence, course, or outcome reported little or no information about total Cr or Cr(VI) concentrations at the workplaces of female chromium production workers [Shmitova 1978, 1980] or male welders that were also spouses [Bonde et al. 1992; Hjollund et al. 1995, 1998, 2000]. The lack of consistent findings and exposure-response analysis precludes formation of conclusions about occupational Cr(VI) exposure and adverse effects on pregnancy and childbirth. Further research is needed.

4.2.4 Other Health Effects

4.2.4.1 Mortality studies

More than 30 studies examined numerous noncancer causes of death in jobs with potential chromium exposure, such as chromate production, chromate pigment production, chromium plating, ferrochromium production, leather tanning, welding, metal polishing, cement finishing, stainless steel grinding or production, gas generation utility work, and paint production or spraying. (Studies previously cited by NIOSH [1975, 1980] are not included.)

Most studies found no statistically significant increases (i.e., $P < 0.05$) in deaths from nonmalignant respiratory diseases, cardiovascular diseases, circulatory diseases, accidents, or any other noncancer cause of death that was included [Hayes et al. 1979, 1989; Korallus et al. 1993; Satoh et al. 1981; Sheffet et al. 1982; Royle 1975a; Franchini et al. 1983; Sorahan and Harrington 2000; Axelsson et al. 1980; Becker et al. 1985; Becker 1999; Blair 1980; Dalager et al. 1980; Järvholm et al. 1982; Silverstein et al. 1981; Sjögren et al. 1987; Svensson et al. 1989; Bertazzi et al. 1981; Blot et al. 2000; Montanaro et al. 1997; Milatou-Smith et al. 1997; Moulin et al. 2000; Pastides et al. 1994a; Simonato et al. 1991; Takahashi and Okubo 1990; Luippold et al. 2005]. However, these studies did not include further investigation of the nonsignificant outcomes and therefore do not confirm the absence of an association.

Some studies did identify significant increases in deaths from various causes [Davies et al. 1991; Alderson et al. 1981; Sorahan et al. 1987; Deschamps et al. 1995; Itoh et al. 1996; Rafnsson and Jóhannesdóttir 1986; Gibb et al. 2000b; Kano et al. 1993; Luippold 2003; Moulin et al. 1993; Rosenman and Stanbury 1996; Stern et al. 1987; Stern 2003]. However, the findings were not consistent: no noncancer cause of death was found to be significantly increased in at least five studies. Furthermore, exposure-response relationships were not examined for those outcomes. Therefore, the results of these studies do not support a causal association between occupational Cr(VI) exposure and a nonmalignant cause of death.

4.2.4.2 Other health effects

NIOSH [1975a] concluded that Cr(VI) exposure could cause other health effects such as kidney damage, liver damage, pulmonary congestion and edema, epigastric pain, and erosion and discoloration of the teeth. Other effects of exposure to chromic acid and chromates not discussed elsewhere in this section include eye injury, leukocytosis, leukopenia, and eosinophilia [NIOSH 2003c; Johansen et al. 1994]. Acute renal failure and acute chromium intoxication occurred in a male worker following a burn with concentrated chromic acid solution to 1% of his body [Stoner et al. 1988].

There has been little post-1975 research of those effects in occupational cohorts. Furthermore, there is insufficient evidence to conclude that occupational exposure to respirable Cr(VI) is related to other health effects infrequently reported in the literature after the NIOSH [1975a] review. These effects included cerebral arachnoiditis in 47 chromium industry workers [Slyusar and Yakovlev 1981] and cases of gastric disturbances (e.g., chronic gastritis, polyps, ulcers, and mucous membrane erosion) in chromium salt workers [Sterekhova et al. 1978]. Neither study analyzed the relationship of air Cr(VI) concentrations and health effects, and one had no comparison group [Sterekhova et al. 1978].

Table 4–1. IARC [1990]-reviewed epidemiologic studies of cancer in workers in chromate-producing industries.

Reference and country	Study population and follow-up	Reference population	Cancer of respiratory organs			Cancer at other sites			Cohort smoking information available and analyzed	Sampling conducted and Cr(VI) identified
			Site	Number of deaths or cases	Estimated relative risk	Site	Number of deaths or cases	Estimated relative risk		
Alderson et al. [1981], United Kingdom	Same U.K. chromate producing factories as Bidstrup and Case [1956]; employed ≥ 1 yr 1948–1977; 2,715 males.	Cancer mortality—England, Wales, Scotland	Lung	116 deaths	2.4*	Other sites Nasal cancer	80 2	1.2 7.1*	No	No
Baetjer [1950], United States	290 male lung cancer patients admitted to two hospitals near U.S. chromate plant 1925–1948.	Random sample of hospital admissions	Lung or bronchi	11 reported exposure to chromium	Reported as statistically significant	—	—	—	No	No
Bidstrup and Case [1956], United Kingdom	Three U.K. chromate factories; mortality follow-up of 723 men employed 1949–1955.	Cancer mortality—England and Wales	Lung	12	3.6*	Other sites	9	1.1	No	No
Brinton et al. [1952], United States	Male workers in seven chromate plants; active employees 1940–1950.	U.S. male mortality, white, nonwhite	Respiratory system, except larynx	10 white; 16 nonwhite	14.3* 80.0*	Other sites	5 white; 1 nonwhite	1.0	No	No

See footnotes at end of table.

(Continued)

Table 4–1 (Continued). IARC [1990]-reviewed epidemiologic studies of cancer in workers in chromate-producing industries.

Reference and country	Study population and follow-up	Reference population	Cancer of respiratory organs			Cancer at other sites			Cohort smoking information available and analyzed	Sampling conducted and Cr(VI) identified
			Site	Number of deaths or cases	Estimated relative risk	Site	Number of deaths or cases	Estimated relative risk		
De Marco et al. [1988], Italy	540 Italian chromate producers employed 1948–1985 with ≥ 1 year cumulative exposure entered into study ≥ 10 years after starting work.	Italian cause-specific death rates	Lung Highly exposed (qualitative estimate of Cr[VI] exposure)	14 6	2.2* 4.2*	Larynx Pleura	3 3	2.9 30.0*	No	No
Federal Security Agency [1953], United States	Health survey of 897 chromate workers in six chromate-producing plants.	Boston chest X-ray survey	Bronchogenic/ Lung	7 white; 3 nonwhite	53.6 (prevalence ratio)	—	—	—	No	Yes
Hayes et al. [1979], United States	2,101 male workers (restricted to 1,803 workers) employed in a U.S. chromate plant ≥ 90 days 1945–1974, working in new and/or old production sites.	Baltimore city mortality	Trachea, bronchus, lung	59	2.0*	Digestive system Other	13 14	0.60 0.40	No	No
Korallus et al. [1982], Germany	1,140 male workers employed > 1 year 1934–1979 at two German chromate plants.	North-Rhine Westphalia mortality	Respiratory organs	51	2.1*	Stomach	12	0.94	No	No

(Continued)

See footnotes at end of table.

Table 4-1 (Continued). IARC [1990]-reviewed epidemiologic studies of cancer in workers in chromate-producing industries.

Reference and country	Study population and follow-up	Reference population	Cancer of respiratory organs			Cancer at other sites			Cohort smoking information available and analyzed	Sampling conducted and Cr(VI) identified
			Site	Number of deaths or cases	Estimated relative risk	Site	Number of deaths or cases	Estimated relative risk		
Machle and Gregorius [1948], United States	Male workers in seven chromate plants; active employees 1930–1947; 193 deaths.	Male oil refinery workers 1933–1938	Respiratory system	42	20.7	Digestive tract	13	2.0	No	Reported as "chromates" (see NIOSH [1975a])
						Oral region (also included in respiratory system)	3	5.4*		
Mancuso and Hueper [1951]; Mancuso [1975], United States	332 U.S. chromate plant workers employed ≥ 1 year 1931–1937; all jobs related to exposure to soluble and insoluble chromium; mortality followed through 1974.	No independent comparison group	Lung	41	—	—	—	—	No	Soluble chromium described as "chiefly hexavalent"

See footnotes at end of table.

(Continued)

Table 4–1 (Continued). IARC [1990]-reviewed epidemiologic studies of cancer in workers in chromate-producing industries.

Reference and country	Study population and follow-up	Reference population	Cancer of respiratory organs			Cancer at other sites			Cohort smoking information available and analyzed	Sampling conducted and Cr(VI) identified
			Site	Number of deaths or cases	Estimated relative risk	Site	Number of deaths or cases	Estimated relative risk		
Satoh et al. [1981], Japan	896 male workers in chromium manufacturing plant in Japan employed ≥ 1 year between 1918 and 1975; mortality followed until 1978, or death. 84% of chromium compounds manufactured 1934–1975 were hexavalent compounds.	Age-, cause-specific mortality, Japanese males	Respiratory cancer Years worked: 1–10 11–20 ≥ 21	31 (includes six sinonasal) 5 9 17	9.2* 4.2* 7.5* 17.5*	Stomach	11	1.0	No	No
Taylor [1966]; Enterline [1974], United States	1,200 males [Enterline 1974] from three U.S. chromate plants, employed 1937–1940 and surveyed 1941–1960.	Cancer mortality; U.S. males 1950, 1953, 1958	Respiratory cancer	69 (2 maxillary sinus)	9.4*	Digestive system	16	1.5	No	No
Watanabe and Fukuchi [1984], Japan	273 chromate production workers employed ≥ 5 years 1947–1973 and followed for mortality 1960–1982.	Age-, year-, cause-specific mortality, Japanese males	Lung	25 (plus one maxillary sinus)	18.3*	Digestive system	6	0.9	No	No

Source: Adapted from IARC [1990].
Dash in "Estimated relative risk" indicates not reported.
*Significant at 95% level.

Table 4–2. IARC [1990]-reviewed epidemiologic studies of cancer in workers in chromate-pigment industries.

Reference and country	Study population and follow-up	Reference population	Cancer of respiratory organs — Site	Number of deaths or cases	Estimated relative risk	Cancer at other sites — Site	Number of deaths or cases	Estimated relative risk	Cohort smoking information available and analyzed	Sampling conducted and Cr(VI) identified
Davies [1978, 1979, 1984], United Kingdom	1,002 male workers in three chromate pigment factories: A, lead and zinc chromate; B, lead and zinc chromate; C, lead chromate; followed up to 1981.	Mortality, England and Wales	Lung; ≥1 year worked, "high" or "medium" exposure to chromate-containing dust: A (entered 1932–1954) B (1948–1967) "high," "medium," or "low" exposure: C (1946–1960)	21 11 7	2.2* 4.4* 1.1	Nasal sinuses Larynx	1 2	5 2.15	No (smoking habits of lung cancer cases reported only)	No
Frentzel-Beyme [1983], Germany, Netherlands	978 male workers from five factories employed > 6 months in three German or Dutch factories manufacturing zinc and lead chromates and followed for 15,076 person-years.	Local death rates for Federal Republic of Germany and the Netherlands	Lung	19	2.0*	—	—	—	No	No
Haguenoer et al. [1981], France	251 male workers in a lead and zinc chromate pigment factory employed > 6 months between 1958 and 1977.	Standard death rates, northern France 1958–1977	Lung	11	4.6*	—	—	—	No (Smoking habits of cancer cases reported only)	No

See footnotes at end of table.

(Continued)

Table 4–2 (Continued). IARC-reviewed epidemiologic studies of cancer in workers in chromate-pigment industries.

Reference and country	Study population and follow-up	Reference population	Cancer of respiratory organs			Cancer at other sites			Cohort smoking information available and analyzed	Sampling conducted and Cr(VI) identified
			Site	Number of deaths or cases	Estimated relative risk	Site	Number of deaths or cases	Estimated relative risk		
Langård and Norseth [1975, 1979]; Langård and Vigander [1983], Norway	133 Norwegian workers producing zinc chromate pigments employed between 1948 and December 1972. Twenty-four workers had more than 3 years of employment to 1972. Cohort was observed to the end of 1980.	Cancer incidence, Norway 1955–1976	Lung	6 (excluding one case with < 3 years' employment)	44	Gastrointestinal	3	6.4	No (Smoking habits of cancer cases reported only)	No and Yes: Exposure reported as μg/m³ or mg/m³ of chromium by Langård and Norseth [1975] and Langård and Vigander [1983]; later reported as mg/m³ of Cr(VI) in a review by Langård [1993].
						Nasal cavity	1	—		
Sheffet et al. [1982]; Hayes et al. [1989], United States	1,181 white and 698 nonwhite males employed in a lead and zinc chromate pigment factory for ≥ 1 month between 1940 and 1969; followed to end of 1982.	Mortality, U.S. white and nonwhite males	Lung ≥ 30 years after initial employment and:	24	1.4	Stomach	6	1.8	No	No
			< 1 year employment	3	1.4†					
			1–9 years' employment	3	2.0†					
			> 10 years' employment	6	3.2†					

Source: Adapted from IARC [1990].
Dash in "Estimated relative risk" indicates not reported.
*Significant at 95% level.
†P for trend < 0.01.

Table 4-3. IARC [1990]-reviewed studies of workers in chromium plating industries.

Reference and country	Study population and follow-up	Reference population	Cancer of respiratory organs			Cancer at other sites			Cohort smoking information available and analyzed	Sampling conducted and Cr(VI) Identified
			Site	Number of deaths or cases	Estimated relative risk	Site	Number of deaths or cases	Estimated relative risk		
Franchini et al. [1983], Italy	178 male workers from nine chrome plating plants (116 in "thick" plating; 62 in "thin") employed ≥ 1 year between 1951 and 1981.	Italy, male mortality	Lung	3	3.3 (4.3* for "thick" platers")	All sites Stomach Pancreas	2 2 2	1.9 4 18*	No	Yes; Chromium trioxide CrO$_3$; 1980 averages: 7 μg/m^3 near plating baths; 3 μg/m^3 in middle of the room.
Okubo and Tsuchiya [1977, 1979, 1987], Japan	Japanese chromium platers; 952 male and female workers with > 6 months' experience. Average follow-up period was 5.2 years for the chromium workers and 5.1 years for controls.	Platers not exposed to chromium; clerical and unskilled workers	Lung	0	—	All sites	5	0.5	No	No
Royle [1975a,b], United Kingdom	Mortality study of 1,056 past and current male platers in 54 chromium-plating plants, employed ≥ 3 months; 130 men had died by May 31, 1974 (Female workers were also studied.)	1,099 non-exposed males in the plants and in two nonplating industries.	Lung and pleura	24	1.4	All sites (including lung) Gastro-intestinal Other sites (excluding lung, gastro-intestinal)	44 8 12	1.7* 1.5 1.9	Yes. Information available; smoking habits of platers were compared with controls—"no important differences."	Yes, at 42 plants. Reported "chromic acid air content" at breathing zone height was generally < 0.03 mg/m^3.

See footnotes at end of table.

(Continued)

Table 4–3 (Continued). IARC [1990]-reviewed studies of workers in chromium plating industries.

Reference and country	Study population and follow-up	Reference population	Cancer of respiratory organs			Cancer at other sites			Cohort smoking information available and analyzed	Sampling conducted and Cr(VI) Identified
			Site	Number of deaths or cases	Estimated relative risk	Site	Number of deaths or cases	Estimated relative risk		
Silverstein et al. [1981], United States	Workers with ≥ 10 years of service in a die-casting and nickel and chrome electroplating plant; 238 deaths (white and nonwhite) between 1974 and 1978.	U.S. national mortality statistics	Lung: White men White women	28 10	1.9* 3.7*	All sites (men) Larynx (men) Stomach (men) Lymphosarcoma, reticulosarcoma (men)	53 2 4 2	1.4* 3.3 2.5 2.9	No	Limited to only a few samples of airborne chromic acid.
Sorahan et al. [1987], United Kingdom	2,689 nickel and chromium platers (1,288 men; 1,401 women). First employed 1946–1975 for ≥ 6 months and observed 1946–1983.	Mortality, England and Wales	Lung, bronchus: Men Women Larynx: Men Women Nose, nasal cavities (men and women)	63 9 3 0 3	1.6* 1.1 3.0 — 10*	Stomach (men and women) Liver Men Women All sites (men and women)	25 4 0 213	1.5 6.7* — 1.3*	No	Yes, as chromic acid. Median value of 60 "measurements" before 1973 was "not detectable or trace." After 1973, majority of measurements were recorded in factory records as "less than 0.05 mg/m³".

Source: Adapted from IARC [1990].
Dash in "Estimated relative risk" indicates not reported.
*Significant at 95% level.

Table 4-4. IARC [1990]-reviewed epidemiologic studies of cancer in workers in ferrochromium industries.

Reference and country	Study population and follow-up	Reference population	Cancer of respiratory organs			Cancer at other sites			Cohort smoking information available and analyzed	Sampling conducted and Cr(VI) identified
			Site	Number of deaths or cases	Estimated relative risk	Site	Number of deaths or cases	Estimated relative risk		
Axelsson et al. [1980], Sweden	1,876 male workers employed ≥ 1 year 1930–1975 in a ferrochromium plant; traced by parish lists and cancer registry.	County deaths, male or national statistics (incidence)	Lung, trachea, bronchus, pleura: All workers Maintenance workers Arc furnace workers	7 4 (2 mesotheliomas) 2 (1 mesothelioma)	1.2 4.0* 1.0	Prostate (all workers)	23	1.2	No	Yes (Cr^{6+} and Cr^{3+}). Cr^{6+} exposures ranged 0–0.25 mg/m³. Sampling method not described.
Langård et al. [1980, 1990], Norway	1,235 male ferrochromium and ferrosilicon workers employed > 1 year 1928–1965 and observed 1953–1985.	General population and internal comparison group	Lung (ferrochromium workers)	10	1.5	All sites (all workers) Ferrochromium workers: Kidney Prostate Stomach	132 5 12 7	0.8 2.8 1.5 1.4	No	Yes, in 1975 survey, mean atmospheric concentration of chromium ranged from 0.01 mg/m³ to 0.29 mg/m³ with a water-soluble content of 11%–33%. Authors stated "Water soluble chromium compounds are considered to be in the hexavalent state."

(Continued)

See footnotes at end of table.

Table 4-4 (Continued). IARC [1990]-reviewed epidemiologic studies of cancer in workers in ferrochromium industries.

Reference and country	Study population and follow-up	Reference population	Cancer of respiratory organs			Cancer at other sites			Cohort smoking information available and analyzed	Sampling conducted and Cr(VI) identified
			Site	Number of deaths or cases	Estimated relative risk	Site	Number of deaths or cases	Estimated relative risk		
Pokrovskaya and Shabynina [1973], USSR	Male and female chromium ferroalloy production workers employed between 1955 and 1969.	Mortality, general population of municipality	Lung (men)	Not reported	4.4 (age 30–39) 6.6* (age 50–59)	All sites (men)	Not reported	3.3* (age 50–59)	No	Yes, specific concentrations and sampling methods not reported—average hexavalent concentrations were 2–7 times greater than allowed.
						Esophagus (men)	Not reported	2.0* (age 50–59) 11.3* (age 60–69)		

Source: Adapted from IARC [1990].
Dash in "Estimated relative risk" indicates not reported.
*Significant at 95% level.

5 Experimental Studies

Experimental studies provide important information about the pharmacokinetics, mechanisms of toxicity, and potential health effects of hexavalent chromium (Cr[VI]) compounds. Studies using cell culture and in vitro techniques, animal models, and human volunteers provide data about these compounds. The results of these experimental studies, when considered with the results of other health effects studies, provide a more comprehensive database for the evaluation of the mechanisms and health effects of occupational exposure to Cr(VI) compounds.

5.1 Pharmacokinetics

The absorption of inhaled Cr(VI) depends on the oxidation state, particle size, and solubility of the compound [ATSDR 2012]. Large particles (> 10 µm) of inhaled Cr(VI) compounds are deposited in the upper respiratory tract; smaller particles can reach the lower respiratory tract. Some of the inhaled Cr(VI) is reduced to trivalent chromium (Cr[III]) in the epithelial or interstitial lining fluids within the bronchial tree. The extracellular reduction of Cr(VI) to Cr(III) reduces the cellular uptake of chromium because Cr(III) compounds cannot enter cells as readily as Cr(VI) compounds. At physiological pH most Cr(VI) compounds are tetrahedral oxyanions that can cross cell membranes. Cr(III) compounds are predominantly octahedral structures to which the cell membrane is practically impermeable. Cr(III) can enter the cell only via pinocytosis [Jennette 1979]. The Cr(VI) ions that cross the cell membrane become a target of intracellular reductants. The Cr(VI) concentration decreases with increasing distance from the point of entry as Cr(VI) is reduced to Cr(III). The Cr(III) ions are transported to the kidneys and excreted.

Inhaled Cr(VI) that is not absorbed in the lungs may enter the gastrointestinal tract following mucociliary clearance. Much of this Cr(VI) is rapidly reduced to Cr(III) by reductants in the saliva and gastric juice and excreted in the feces. The remaining 3% to 10% of the Cr(VI) is absorbed from the intestines into the blood stream, distributed throughout the body, transported to the kidneys, and excreted in the urine [Costa 1997; Weber 1983].

5.2 Mechanisms of Toxicity

The possible mechanisms of the genotoxicity and carcinogenicity of Cr(VI) compounds have been reviewed [Holmes et al. 2010; Nickens et al. 2010]. However, the exact mechanisms of Cr(VI) toxicity and carcinogenicity are not yet fully understood. A significant body of research suggests that Cr(VI) carcinogenicity may result from damage mediated by the bioreactive products of Cr(VI) reduction, which include the Cr(VI) intermediates (Cr[V] and Cr[IV]), and reactive oxygen species (ROS). Factors that may affect the toxicity of a chromium compound include its bioavailability, oxidative properties, and solubility [Långard 1993; Katz and Salem 1993; De Flora et al. 1990; Luo et al. 1996; Klein et al. 1991].

Intracellular Cr(VI) undergoes metabolic reduction to Cr(III) in microsomes, in mitochondria, and by cellular reductants such as ascorbic acid, lipoic acid, glutathione, cysteine, reduced nicotinamide adenine dinucleotide phosphate (NADPH), ribose, fructose, arabinose and diol- and thiol-containing molecules, as well as NADPH/flavoenzymes. Although the extracellular reduction of Cr(VI) to Cr(III) is a mechanism of detoxification as it decreases the number of bioavailable Cr(VI) ions, intracellular reduction may be an essential element in the mechanism of intracellular Cr(VI) toxicity.

The intracellular Cr(VI) reduction process generates products including Cr(V), Cr(IV), Cr(III) molecular oxygen radicals, and other free radicals. The molecular oxygen is reduced to superoxide radical, which is further reduced to hydrogen peroxide (H_2O_2) by superoxide dismutase (SOD). H_2O_2 reacts with Cr(V), Cr(IV) or Cr(III) to generate hydroxyl radicals (˙OH) via the Fenton-like reaction, and it undergoes reduction-oxidation cycling [Ding and Shi 2002]. The high concentration of oxygen radicals and other free radical species generated in the process of Cr(VI) reduction may result in a variety of lesions on nuclear chromatin, leading to mutation and possible neoplastic transformation [Kasprzak 1991].

In the presence of cellular reducing systems that generate chromium intermediates and hydroxyl radicals, Cr(VI) salts induce various types of DNA damage, resulting either from the breakage of existing covalent bonds or the formation of new covalent bonds among molecules, such as DNA interstrand crosslinks, DNA-protein crosslinking, DNA double strand breaks, and depurination. Such lesions could lead to mutagenesis and ultimately to carcinogenicity [Shi et al. 1994; Tsapakos and Wetterhahn 1983; Tsapakos et al. 1983; Sterns et al. 1995; Sugiyama et al. 1986; Singh et al. 1998; Ding and Shi 2002; Fornace et al. 1981]. The oxidative damage may result from a direct binding of the reactive Cr(VI) intermediates to the DNA or may be due to the indirect effect of ROS interactions with nuclear chromatin, depending on their intracellular location and proximity to DNA [Ding and Shi 2002; Shi and Dalal 1990a,b,c; Singh et al. 1998; Liu et al. 1997b]. Cr(VI) does not bind irreversibly to native DNA and does not produce DNA lesions in the absence of the microsomal reducing systems in vitro [Tsapakos and Wetterhahn 1983].

In addition to their oxidative properties, the solubility of Cr(VI) compounds is another important factor in the mechanism of their carcinogenicity. Animal studies indicate that insoluble and sparingly soluble Cr(VI) compounds may be more carcinogenic than soluble chromium compounds [Levy et al. 1986].

Particles of lead chromate, a relatively insoluble Cr(VI) compound, when added directly to the media of mammalian cell culture, induced cell transformation [Douglas et al. 1980]. When injected into whole animals, the particles produced tumors at the site of injection [Furst et al. 1976]. Several hypotheses have been proposed to explain the effects of insoluble Cr(VI) compounds. One hypothesis proposes that particles dissolve extracellularly, resulting in chronic, localized exposure to ionic chromate. This hypothesis is consistent with studies demonstrating that extracellular dissolution is required for lead chromate-induced clastogenesis [Wise et al. 1993, 1994; Xie et al. 2004]. Xie et al. [2004] demonstrated that lead chromate clastogenesis in human bronchial cells is mediated by the extracellular dissolution of the particles but not their internalization.

Another hypothesis suggests that a high Cr(VI) concentration is created locally inside the cell during internalization of Cr(VI) salt particles by phagocytosis [Leonard et al. 2004].

High intracellular local Cr(VI) concentrations can generate high concentration of ROS inside the cell, which may overwhelm the local ROS scavenging system and result in cytotoxicity and genotoxicity [Kasprzak 1991]. Highly soluble compounds do not generate such high local concentrations of Cr(VI). However, once inside the cell, both soluble (sodium chromate) and insoluble (lead chromate) Cr(VI) compounds induce similar amounts and types of concentration-dependent chromosomal damage in exposed cultured mammalian cells [Wise et al. 1993, 2002, 2003]. Pretreatment of these cells with ROS scavengers such as vitamin E or C prevented the toxic effects of both sodium chromate and lead chromate.

Numerous studies report a broad spectrum of cellular responses induced by exposure to various Cr(VI) compounds. These cytotoxic and genotoxic responses are consistent with mechanistic events associated with carcinogenesis. Studies in human lung cells provide data regarding the genotoxicity of many Cr(VI) compounds [Wise et al. 2006a; Holmes et al. 2006b; Xie et al. 2009]. Cr(VI) compounds induce transformation of human cells, including bronchial epithelial cells [Xie et al. 2007; Xie et al. 2008].

Barium chromate induced concentration-dependent chromosomal damage, including chromatid and chromosomal lesions, in human lung cells after 24 hours of exposure [Wise et al. 2003]. Lead chromate and soluble sodium chromate induced concentration-dependent chromosomal aberration in human bronchial fibroblasts after 24 hours of exposure [Wise et al. 2002; Xie et al. 2004]. Cotreatment of cells with vitamin C blocked the chromate induced toxicity. Calcium chromate induced DNA single-strand breaks and DNA protein cross-links in a dose-dependent manner in three cell lines. Exposing human lung cell cultures to lead chromate induced chromosome instability including centrosome amplification and aneuploidy [Holmes et al. 2006a] and spindle assembly checkpoint bypass [Wise et al. 2006b]. Sodium dichromate generated ROS that increased the level and activity of the protein p53 in human lung epithelial cells. In normal cells the protein p53 is usually inactive. It is usually activated to protect cells from tumorigenic alterations in response to oxidative stress and other stimuli such as ultraviolet or gamma radiation. An increased ·OH concentration activated p53; elimination of ·OH by H_2O_2 scavengers inhibited p53 activation [Ye et al. 1999; Wang et al. 2000; Wang and Shi 2001].

The ROS (mainly H_2O_2) formed during potassium chromate reduction induced the expression of vascular endothelial growth factor (VEGF) and hypoxia-induced factor 1 (HIF–1) in DU145 human prostate carcinoma cells. VEGF is the essential protein for tumor angiogenesis. HIF–1, a transcription factor, regulates the expression of many genes including VEGF. The level of HIF–1 activity in cells correlates with the tumorigenic response and angiogenesis in nude mice, is induced by the expression of various oncogenes, and is overexpressed in many human cancers [Gao et al. 2002; Ding and Shi 2002].

Early stages of apoptosis have been induced in human lung epithelial cells in vitro following exposure to potassium dichromate. Scavengers of ROS, such as catalase, aspirin, and N-acetyl-L-cysteine, decreased apoptosis induced by Cr(VI); reductants such as NADPH and glutathione enhanced it. Apoptosis can be triggered by oxidative stress. Agents that promote or suppress apoptosis may change the rates of cell division and lead to the neoplastic transformation of cells [Singh et al. 1998; Ye et al. 1999; Chen et al. 1999].

The treatment of mouse macrophage cells in vitro with sodium chromate induced a dose-dependent activation of the transcription enhancement factors NF-κB and AP-1 [Chen et al. 1999, 2000]. Activation of these factors represents a primary cellular oxidative stress response. These factors enhance the transcription of many genes and the enhanced expression of oncogenes [Ji et al. 1994].

Sodium dichromate increased tyrosine phosphorylation in human epithelial cells. The phosphorylation could be inhibited by antioxidants [Wang and Shi 2001]. Tyrosine phosphorylation is essential in the regulation of many cellular functions, including cancer development [Qian et al. 2001].

Human lung epithelial A549 cells exposed to potassium dichromate in vitro generated ROS-induced cell arrest at the G2/M phase of the cell proliferation cycle at relatively low concentrations and apoptosis at high concentrations. Interruption of the proliferation process is usually induced in response to cell damage, particularly DNA damage. The cell remains arrested in a specific cell cycle phase until the damage is repaired. If damage is not repaired, mutations and cell death or cancer may result [Zhang et al. 2001].

Gene expression profiles indicate that exposing human lung epithelial cells to potassium dichromate in vitro resulted in up regulation of the expression of 150 genes, and down regulation of 70 genes. The analysis of gene expression profiles indicated that exposure to Cr(VI) may be associated with cellular oxidative stress, protein synthesis, cell cycle regulation, and oncogenesis [Ye and Shi 2001].

These in vitro studies have limitations of models of human exposure because they cannot account for the detoxification mechanisms that take place in intact physiological systems. However, these studies represent a body of data on cellular responses to Cr(VI) that provide important information regarding the potential genotoxic mechanisms of Cr(VI) compounds. The cellular damage induced by these compounds is consistent with the mechanisms of carcinogenesis.

5.3 Health Effects in Animals

Chronic inhalation studies provide the best data for extrapolation to airborne occupational exposure. Only a few of these chronic inhalation studies have been conducted using Cr(VI) compounds. Glaser et al. [1985, 1990] conducted subchronic inhalation studies of sodium dichromate exposure in rats. Adachi et al. [1986, 1987] conducted chronic inhalation studies of chromic acid mist exposure in mice. Glaser et al. [1986] conducted chronic inhalation studies of sodium dichromate exposure in rats. Steinhoff et al. [1986] conducted an intratracheal study of sodium dichromate exposure in rats. Levy et al. [1986] conducted an intrabronchial implantation study of various Cr(VI) materials in rats. The results of these animal studies support the classification of Cr(VI) compounds as occupational carcinogens.

5.3.1 Subchronic Inhalation Studies

Glaser et al. [1985] exposed male Wistar rats to whole body aerosol exposures of sodium dichromate at 0, 25, 50, 100, or 200 µg Cr(VI)/m^3 for 22 hr/day, 7 days/wk for 28 or 90 days. Twenty rats were exposed at each dose level. An additional 10 rats were exposed at 50 µg for 90 days followed by 2 months of nonexposure before sacrifice. The average mass median diameter (MMD) of the aerosol particles was 0.2 µm. Significant increases ($P < 0.05$) occurred in the serum triglyceride, phospholipid contents, and mitogen-stimulated splenic mean T-lymphocyte count of rats exposed at the 200 µg/m³ level for 90 days. Serum total immunoglobulins were statistically increased ($P < 0.01$) for the 50 and 100 µg exposure groups.

To further study the humoral immune effects, half of the rats in each group were immunized with sheep red blood cells 4 days before sacrifice [Glaser et al. 1985]. The primary antibody responses for IgM B-lymphocytes were statistically increased ($P < 0.05$) for the groups exposed to 25 µg Cr(VI)/m³ and higher. The mitogen-stimulated T-lymphocyte response of spleen cells to Concanavalin A was significantly increased ($P < 0.05$) for the 90-day, 200 µg/m³ group compared with the control group. The mean macrophage cell counts were significantly lower ($P < 0.05$) than control values for only the 50 and 200 µg Cr(VI)/m³, 90-day groups. Alveolar macrophage phagocytosis was statistically increased in the 50 µg level of the 28-day study, and the 25 and 50 µg mg/m³ Cr(VI) levels of the 90-day study ($P < 0.001$). A significant depression of phagocytosis occurred in the 200 µg/m³ group of the 90-day study versus controls.

A group of rats exposed to 200 µg Cr(VI)/m³ for 42 days and controls received an acute iron oxide particulate challenge to study lung clearance rates during a 49-day nonexposure post-challenge period [Glaser et al. 1985]. Iron oxide clearance was dramatically and increasingly decreased in a bi-exponential manner for the group exposed to Cr(VI) compared with the controls.

Glaser et al. [1990] studied lung toxicity in animals exposed to sodium dichromate aerosols. Groups of 30 male Wistar rats were exposed to 0, 50, 100, 200, or 400 µg Cr(VI)/m³ for 22 hr/day, 7 days/week for 30 or 90 days, followed by a 30-day nonexposure recovery period. Aerosol mass median aerodynamic diameter (MMAD) ranged from 0.28 to 0.39 µm. Sacrifices of 10 rats occurred after experimental days 30, 90, and 120. The only sign or symptom induced was an obstructive dyspnea present at the 200 and 400 µg/m³ levels. Statistically significant reductions in body weight gains were present at 30 days for the 200 µg level, with similar reductions for the 400 µg level rats at the 30, 90, and 120-day intervals. White blood cell counts were statistically increased ($P < 0.05$) for all four dichromate exposure groups for the 30- and 90-day intervals, but the white blood cell counts returned to control levels after 30 days of nonexposure. The lung parameters studied had statistically significant dose-related increases after either 30 or 90 days of inhalation exposure to dichromate; some remained elevated despite the nonexposure recovery period. A No Observed Adverse Effect Level (NOAEL) was not achieved.

Bronchoalveolar lavage (BAL) provided information about pulmonary irritation induced by sodium dichromate exposure in these rats [Glaser et al. 1990]. Total protein levels present on day 30 progressively decreased at days 90 and 120 but remained above control values. Alveolar vascular integrity was compromised as BAL albumin levels were increased for all treatment groups, with only the 200 and 400 µg/m³ levels remaining above those of the controls at the end of the recovery period. Lung cell cytotoxicity as measured by cytosolic lactate dehydrogenase, and lysosomal ß-glucuronidase was increased by dichromate exposure but normalized during the post-exposure period. Mononuclear macrophages comprised 90% of recovered total BAL cells. The two highest exposure groups had equal increases throughout the treatment period, but they returned to normal during the recovery period. These macrophages had higher cell division rates, sometimes were multinuclear, and were bigger when compared with control cells. Sodium dichromate exposure induced statistically significant increased lung weights for the 100, 200, and 400 µg/m³ groups throughout the study, including the nonexposure period. Histopathology of lung tissue revealed an initial bronchoalveolar hyperplasia for all exposure groups at day 30, while only the 200 and 400 µg/m³ levels retained some lower levels of hyperplasia at study day 120. There was also an initial lung fibrosis observed in some animals at the levels above 50 µg/m³ on day 30, which was not present during the remainder of the study. Lung histiocytosis remained elevated throughout the entire study for all treatment groups.

5.3.2 Chronic Inhalation Studies

Adachi et al. [1986] exposed 50 female ICR/JcI mice to 3.63 mg Cr(VI)/m³ chromic acid mist (85% of mist measuring < 5 µm) for 30 min/day, 2 days/week for 12 months, followed by a 6-month nonexposure recovery period. Proliferative changes were observed within the respiratory tract after 26 weeks of chromate exposure. Pin-hole-sized perforations of the nasal septum occurred after 39 weeks at this exposure level. When the incidence rates for histopathological findings (listed below) for chromate-exposed animals were compared for successive study periods, the treatment group data were generally similar for weeks 40–61 when compared with weeks 62–78, with the exception of the induction of two adenocarcinomas of the lungs present in two females at the terminal 78-week sacrifice. The total study pathology incidence rates for the 48 chromate-exposed females were the following: perforated nasal septum (n = 6), tracheal (n = 43)/bronchial (n = 19) epithelial proliferation, and emphysema (n = 11), adenomatous metaplasia (n = 3), adenoma (n = 5), and adenocarcinoma (n = 2) of the lungs. Total control incidence rates for the 20 females examined were confined to the lung: emphysema (n = 1), adenomatous metaplasia (n = 1), and adenoma (n = 2).

Adachi [1987] exposed 43 female C57BL mice to 1.81 mg Cr(VI)/m³ chromic acid mist (with 85% of mist measuring ~5 µm) for 120 min/day, 2 days/week for 12 months, followed by a 6-month nonexposure recovery period. Twenty-three animals were sacrificed at 12 months, with the following nontumorigenic histological changes observed: nasal cavity perforation (n = 3); tracheal hyperplasia (n = 1); and emphysema (n = 9) and adenomatous metaplasia (n = 4) of the lungs. A terminal sacrifice of the 20 remaining females occurred at 18 months, which demonstrated perforated nasal septa (n = 3) and papillomas (n = 6); laryngeal/tracheal hyperplasia (n = 4); and emphysema (n = 11), adenomatous metaplasia (n = 5), and adenoma (n = 1) of the lungs. Only emphysema (n = 2) and lung metaplasia (n = 1) were observed in control females sacrificed after week 78.

Glaser et al. [1986] exposed groups of 20 male Wistar rats to aerosols of 25, 50, or 102 µg/m³ sodium dichromate for 22 to 23 hr/day, 7 days/week for 18 months, followed by a 12-month nonexposure recovery period. Mass median diameter of the sodium dichromate aerosol was 0.36 µm. No clinical sign of irritation induced by Cr(VI) was observed in any treated animal. Statistically increased liver weights (+26%) were observed at 30 months for the 102 µg/m³ dichromate males. Weak accumulations of pigment-loaded macrophages were present in the lungs of rats exposed to 25 µg/m³ sodium dichromate; moderate accumulations were present in rats exposed to 50 and 102 µg/m³ sodium dichromate. Three primary lung tumors occurred in the 102 µg Cr(VI)/m³ group: two adenomas and one adenocarcinoma. The authors concluded that the 102 µg Cr(VI)/m³ level of sodium dichromate induced a weak lung carcinogenic effect in rats exposed under these conditions.

5.3.3 Intratracheal Studies

Steinhoff et al. [1986] dosed Sprague-Dawley rats via intratracheal instillation with equal total weekly doses of sodium dichromate for 30 months: either five consecutive daily doses of 0.01, 0.05, or 0.25 mg/kg or one weekly dose of 0.05, 0.25, or 1.25 mg/kg. Each group consisted of 40 male and 40 female rats. Groups left untreated or given saline were negative controls. Body weight gains were suppressed in males treated with single instillations of 1.25 mg/kg of sodium dichromate. Chromate-induced nonneoplastic and neoplastic lesions were detected only in the lungs. The nonneoplastic pulmonary lesions were primarily found at the maximum tolerated irritant concentration level for the high dose sodium dichromate group rather than having been dependent upon the total dose administered. The nonneoplastic pulmonary lesions occurred predominantly in the highest dose group and were characterized by fibrotic regions that contained residual distorted bronchiolar lumen or cellular inflammatory foci containing alveolar macrophages, proliferated epithelium, and chronic inflammatory thickening of the alveolar septa plus atelectasis. The neoplastic lesions were non-fatal lung tumors found in these chromate-treated animals. Fourteen rats given single weekly instillations of 1.25 mg sodium dichromate/kg developed a significant ($P < 0.01$) number of tumors: 12 benign bronchioalveolar adenomas and 8 malignant tumors including 2 bronchioalveolar adenocarcinomas and 6 squamous cell carcinomas. Only one additional tumor, a bronchioalveolar adenocarcinoma, was found in a rat that had received single weekly instillations of 0.25 mg/kg sodium dichromate.

5.3.4 Intrabronchial Studies

Levy et al. [1986] conducted a 2-year intrabronchial implantation study of 20 chromium-containing materials in Porton-Wistar rats. Test groups consisted of 100 animals with equal numbers of male and female rats. A small, hook-equipped stainless steel wire mesh basket containing 2 mg of cholesterol and test material was inserted into the left bronchus of each animal. Two positive control groups received pellets loaded with 20-methylcholanthrene or calcium chromate. The negative control group received a blank pellet loaded with cholesterol. Pulmonary histopathology was the primary parameter studied. There were inflammatory and metaplastic changes present in the lungs and bronchus, with a high level of bronchial irritation induced by the presence of the basket alone. A total of 172 tumors were obtained throughout the study, with only 18 found at the terminal sacrifice. Nearly all tumors were large bronchial keratinizing squamous cell carcinomas that affected a major part of the left lung and were the cause of death for most affected animals. The authors noted that no squamous cell carcinomas had been found in 500 of their historical laboratory controls.

In Table 5–1, study data from Levi et al. [1986] were transformed by NIOSH to present the rank order of tumor induction potential for the test compounds through calculation of the mean µg of Cr(VI) required to induce a single bronchiolar squamous cell carcinoma. The rank order of tumor induction potential for the positive Cr(VI) compounds using these data was the following: strontium > calcium > zinc > lead, chromic acid > sodium dichromate > barium. The role solubility played in tumor production for these test materials was inconsistent and not able to be determined.

5.3.5 Chronic Oral Studies

The National Toxicology Program (NTP) conducted 2-year drinking water studies of sodium dichromate dehydrate (SDD) in rodents [NTP 2008; Stout et al. 2009]. Male and female F344/N

Table 5–1. Single intrabronchiolar pellet implantation of Cr(VI) or Cr(III) materials and their potential to induce lung carcinomas during a 2-year period in rats

Test compound	Water solubility, mg Cr(VI)/L	Cr(VI) (%)	Pellet Cr(VI) content (µg)	Number of carcinomas	µg Cr(VI) to induce carcinoma*
Strontium chromate	207,000	8.7	174	43	4
Strontium chromate	63,000	24.3	486	62	8
Hi Lime Residue (2.7% calcium chromate)	1,820	1.2	24	1	24
Calcium chromate Positive control	181,000	32.5	649	25	26
Zinc chromate	420	8.7	173	5	35
Zinc chromate	64,000	9.2	184	3	61
Kiln frit†	84,600	9.3	186	2	93
LD chrome yellow supra‡	< 1	5.7	114	1	114
Lead chromate	17	5.7	115	1	115
Vanadium solids/leach†	54,000	7.3	146	1	146
Zinc tetroxychromate	230	8.8	176	1	176
Chromic acid	400,000	21.2	424	2	212
Primrose chrome yellow‡	5	12.6	252	1	252
Med chrome yellow‡	2	16.3	326	1	326
Sodium dichromate dehydrate	328,000	34.8	696	1	696
Molybdate chrome orange‡	< 1	12.9	258	0	—
Light chrome yellow‡	1	12.5	250	—	—
Med chrome yellow‡	17	10.5	210	—	—
Barium chromate	11	6.8	135	0	—
Recycled residue	6,000	0.7	14	0	—
High silica Cr(III) ore	5	13.7	750	0	—
Cholesterol Negative control§	Not reported	NA	NA	0	NA
3-Methylcholanthrene Positive control	Not reported	NA	NA	22¶	NA

Data source: Levy et al. [1986]; calculations by NIOSH.
Abbreviations: NA = Not applicable.
*µg Cr(VI) to induce carcinomas = capsule Cr(VI) content / number of carcinomas
†This process material contained unstated amounts of calcium chromate.
‡Identified also as being a lead chromate containing group.
§No lung tumors were previously found in 500 negative historical control rats that had basket implants.
¶21 squamous cell carcinomas and one anaplastic carcinoma of the lung.

rats and female B6C3F1 mice were exposed to 0, 14.3, 57.3, 172, or 516 mg/L SDD, and male mice to 0, 14.3, 28.6, 85.7, or 257.4 mg/L SDD. Statistically significant concentration-related increased incidences of neoplasms of the oral cavity in male and female rats, and of the small intestine in male and female mice, were reported. The NTP concluded that these results provide clear evidence of the carcinogenic activity of SDD in rats and mice [Stout et al. 2009]. This conclusion was reinforced by the similar results reported between the sexes in both rats and mice.

5.3.6 Reproductive Studies

Reviews and analyses of the animal studies of the reproductive effects of Cr(VI) compounds are available [EPA 1998; ATSDR 2012; Health Council of the Netherlands 2001; 71 Fed. Reg. 10099 (2006); OEHHA 2009]. Animal studies conducted using the oral route of exposure have reported adverse effects on reproductive organs, sperm, fertility, reproductive outcomes, and other adverse reproductive and developmental effects [OEHHA 2009]. Negative studies have also been reported; potassium dichromate administered in the diet to mice and rats did not result in adverse reproductive effects or outcomes in rats or mice [NTP 1996a,b; 1997]. Inhalation studies in male and female rats did not result in adverse reproductive effects [Glaser et al. 1985, 1986, 1988; ATSDR 2012].

The California Environmental Protection Agency (EPA) Office of Environmental Health Hazard Assessment (OEHHA) determined that Cr(VI) compounds "have been clearly shown through scientifically valid testing according to generally accepted principles to cause developmental, male reproductive toxicity, and female reproductive toxicity" [OEHHA 2009]. The Health Council of the Netherlands' Committee for Compounds Toxic to Reproduction recommended that Cr(VI) compounds be classified as "substances which cause concern for human fertility" (category 3) and "substances which should be regarded as if they impair fertility in humans" (category 2) [Health Council of the Netherlands 2001]. OSHA concluded in its evaluation of reproductive studies that there is insufficient evidence to classify Cr(VI) compounds as a reproductive toxin in normal working situations [71 Fed. Reg. 10099 (2006)].

5.4 Dermal Studies

Dermal exposure is another important route of exposure to Cr(VI) compounds in the workplace. Experimental studies have been conducted using human volunteers, animal models, and *in vitro* systems to investigate the dermal effects of Cr(VI) compounds.

5.4.1 Human Dermal Studies

Mali et al. [1963] reported the permeation of intact epidermis by potassium dichromate in human volunteers in vivo. Sensitization was reported in humans exposed to this Cr(VI) compound but not Cr(III) sulfate.

Baranowska-Dutkiewicz [1981] conducted 27 Cr(VI) absorption experiments on seven human volunteers. The forearm skin absorption rate for 0.01 molar solution of sodium chromate was 1.1 µg/cm^2/hr, for 0.1 molar solution it was 6.5 µg/cm^2/hr, and for 0.2 molar solution it was 10.0 µg/cm^2/hr. The amount of Cr(VI) absorbed as a percent of the applied dose decreased with increasing concentration. The absorption rate increased as the Cr(VI) concentration applied increased, and it decreased as the exposure time increased.

Corbett et al. [1997] immersed four human volunteers below the shoulders in water containing 22 mg/L potassium dichromate for

3 hours to assess their uptake and elimination of chromium. The concentration of Cr in the urine was used as the measure of systemic uptake. The total Cr excretion above historical background ranged from 1.4 to 17.5 µg. The dermal uptake rates ranged from approximately 3.3×10^{-5} to 4.1×10^{-4} µg/cm²/hr, with an average of 1.5×10^{-4}. One subject had a dermal uptake rate approximately seven times higher than the average for the other three subjects.

5.4.2 Animal Dermal Studies

Mali et al. [1963] demonstrated the experimental sensitization of 13 of 15 guinea pigs by injecting them with 0.5 mg potassium dichromate in Freund adjuvant subdermally twice at 1-week intervals.

Gad et al. [1986] conducted standard dermal LD_{50} tests to evaluate the acute toxicity of sodium chromate, sodium dichromate, potassium dichromate, and ammonium dichromate salts in New Zealand white rabbits. All salts were tested at 1.0, 1.5, and 2.0 g/kg dosage, with the exception of sodium chromate, which was tested at the two higher doses only. In males, the dermal LD_{50} ranged from a mean of 0.96 g/kg (SD = 0.19) for sodium dichromate to 1.86 g/kg (SD = 0.35) for ammonium dichromate. In females the dermal LD_{50} ranged from a mean of 1.03 g/kg (SD = 0.15) for sodium dichromate to 1.73 g/kg (SD = 0.28) for sodium chromate. Each of the four salts, when moistened with saline and occluded to the skin for 4 hours, caused marked irritation. Occlusion of each salt on the skin of the rabbit's back for 24 hours caused irreversible cutaneous damage.

Liu et al. [1997a] demonstrated the reduction of an aqueous solution of sodium dichromate to Cr(V) on the skin of Wistar rats using in vivo electron paramagnetic resonance spectroscopy. Removal of the stratum corneum by stripping the skin with surgical tape 10 times before the application of the dichromate solution increased the rates of formation and decay of Cr(V).

5.4.3 In Vitro Dermal Studies

Gammelgaard et al. [1992] conducted chromium permeation studies on full thickness human skin in an in vitro diffusion cell system. Application of 0.034 M potassium chromate to the skin resulted in significantly higher levels of chromium in the epidermis and dermis, compared with Cr(III) nitrate and Cr(III) chloride. Chromium levels in the epidermis and dermis increased with the application of increasing concentrations of potassium chromate up to 0.034 M Cr. Chromium skin levels increased with the application of potassium chromate solutions with increasing pH. The percentage of Cr(VI) converted to Cr(III) in the skin was largest at low total chromium concentrations and decreased with increasing total concentrations, indicating a limited ability of the skin to reduce Cr(VI).

Van Lierde et al. [2006] conducted chromium permeation studies on human and porcine skin using a Franz static diffusion cell. Potassium dichromate was determined to permeate human and pig skin after 168 hours of exposure, while the Cr(III) compounds tested did not. Exposure of the skin to 5% potassium dichromate resulted in an increased, but not proportionally increased, amount of total Cr concentration in the skin, compared with exposure to 0.25% potassium dichromate. Exposure to 5% potassium dichromate compared with 2.5% potassium did not result in much more of an increased Cr skin concentration dichromate, indicating a possible limited binding capacity of the skin. A smaller amount of Cr was bound to the skin when the salts were incubated in simulated sweat before application onto the skin. A larger accumulation of Cr was found in the skin after exposure to potassium dichromate compared with Cr(III) compounds.

Rudolf et al. [2005] reported a pronounced effect of potassium chromate on the morphology and motile activity of human dermal fibroblasts at concentrations ranging from 1.5 to 45 µM in tissue culture studies. A time and concentration-dependent effect on cell shrinkage, reorganization of the cytoskeleton, and inhibition of fibroblast motile activity was reported. The inhibitory effect on fibroblast migration was seen at all concentrations 8 hours after treatment; effects at higher doses were seen by 4 hours after treatment. Cr(VI) exposure also resulted in oxidative stress, alteration of mitochondrial function, and mitochondria-dependent apoptosis in dermal fibroblasts.

5.5 Summary of Animal Studies

Cr(VI) compounds have been tested in animals using many different experimental conditions and exposure routes. Although experimental conditions are often different from occupational exposures, these studies provide data to assess the carcinogenicity of the test compounds. Chronic inhalation studies provide the best data for extrapolation to occupational exposure; few have been conducted using Cr(VI) compounds. However, the body of animal studies supports the classification of Cr(VI) compounds as occupational carcinogens.

The few chronic inhalation studies available demonstrate the carcinogenic effects of Cr(VI) compounds in mice and rats [Adachi et al. 1986, 1987; Glaser et al. 1986]. Animal studies conducted using other respiratory routes of administration have also produced positive results with some Cr(VI) compounds. Zinc chromate and calcium chromate produced a statistically significant ($P < 0.05$) number of bronchial carcinomas when administered via an intrabronchial pellet implantation system [Levy et al. 1986]. Cr(VI) compounds with a range of solubilities were tested using this system. Although soluble Cr(VI) compounds did produce tumors, these results were not statistically significant. Some lead chromate compounds produced squamous carcinomas, which although not statistically significant may be biologically significant because of the historical absence of this cancer in control rats.

Steinhoff et al. [1986] administered the same total dose of sodium dichromate either once per week or five times per week to rats via intratracheal instillation. No increased incidence of lung tumors was observed in animals dosed five times weekly. However, in animals dosed once per week, a statistically significant ($P < 0.01$) tumor incidence was reported in the 1.25 mg/kg exposure group. This study demonstrates a dose-rate effect within the constraints of the experimental design. It suggests that limiting exposure to high Cr(VI) levels may be important in reducing carcinogenicity. However, quantitative extrapolation of these animal data to the human exposure scenario is difficult.

Animal studies conducted using nonrespiratory routes of administration have also produced positive results with some Cr(VI) compounds [Hueper 1961; Furst 1976]. These studies provide another data set for hazard identification.

IARC [2012] concluded "there is sufficient evidence in experimental animals for the carcinogenicity of chromium (VI) compounds".

Molecular toxicology studies provide support for classifying Cr(VI) compounds as occupational carcinogens. They demonstrate the cytotoxic and genotoxic effects associated with carcinogenesis of Cr(VI) compounds.

Animal studies using the oral route of exposure demonstrate the carcinogenicity and potential adverse reproductive and developmental toxicity of Cr(VI) compounds. These studies and the nonmalignant health effects of Cr(VI) have been reviewed and evaluated by other government agencies [71 Fed. Reg. 10099 (2006); ATSDR 2012; EPA 1998; Health Council of the Netherlands 2001; NTP 2011; OEHHA 2009].

6 Quantitative Assessment of Risk

6.1 Overview

This chapter uses standard epidemiological and risk assessment methods to present numerical estimates of the excess lifetime risk of lung cancer due to occupational exposure to hexavalent chromium (Cr[VI]) compounds. The excess lifetime risk is the increase in risk over a lifetime above the background level. These estimates are based on mathematical models that describe the relationship between exposure to Cr(VI) and lung cancer deaths in known populations of exposed workers. Exposure-response modeling requires making assumptions about workers' exposures over the course of their working lifetime, and the mathematical form of the exposure-response relationship. A range of excess working lifetime risk estimates is provided that are associated with various levels of Cr(VI) workplace exposure.

Exposure-response data are needed to quantify the risk of occupational exposure to Cr(VI) compounds. The exposure and health data from two chromate production facilities provide the bases for the quantitative risk assessments of lung cancer due to occupational Cr(VI) exposure. Dose-response data from the Painesville, Ohio chromate production facility were analyzed by Crump et al. [2003], K.S. Crump [1995], Gibb et al. [1986], and EPA [1984]. Dose-response data from the Baltimore, Maryland chromium chemical production facility were analyzed by Park et al. [2004], K.S. Crump [1995], and Gibb et al. [1986]. The epidemiologic studies of these worker populations are described in the human health effects chapter (see Chapter 4). Goldbohm et al. [2006] discusses the framework necessary to conduct quantitative risk assessments based on epidemiological studies in a structured, transparent, and reproducible manner.

The Baltimore and Painesville cohorts [Gibb et al. 2000b; Luippold et al. 2003] are the best studies for predicting Cr(VI) cancer risks because of the quality of the exposure estimation, large amount of worker data available for analysis, extent of exposure, and years of follow-up [NIOSH 2005a]. NIOSH selected the Baltimore cohort [Gibb et al. 2000b] for analysis because it had the greater number of lung cancer deaths, better smoking histories, and a more comprehensive retrospective exposure archive.

The results of these occupational quantitative risk assessments demonstrated an elevated risk of lung cancer death to workers exposed to Cr(VI) at the previous NIOSH REL (1 µg Cr(VI)/m^3) over a working lifetime. A risk assessment conducted on the Painesville data reports an excess lifetime risk estimate of lung cancer death of 2 per 1,000 workers at the previous NIOSH REL [Crump et al. 2003]. The NIOSH risk assessment conducted on the Baltimore data indicates an excess lifetime risk estimate of lung cancer death of 6 per 1,000 workers at 1 µg/m^3 and approximately 1 per 1,000 workers at 0.2 µg/m^3 [Park et al. 2004]. These estimates of increased lung cancer risk vary depending on the data set used, the assumptions made, and the models tested.

Environmental risk assessments of Cr(VI) exposure have also been conducted [ATSDR 2012; EPA 1998, 1999]. These analyses assess the risks of nonoccupational Cr(VI) exposure.

6.2 Baltimore Chromate Production Risk Assessments

NIOSH calculated estimates of excess lifetime risk of lung cancer death resulting from occupational exposure to chromium-containing mists and dusts [Park et al. 2004] using data from a cohort of chromate chemical production workers [Gibb et al. 2000b]. NIOSH determined that Gibb et al. [2000b] was the best data set available for quantitative risk assessment because of its extensive exposure assessment and smoking information, strong statistical power, and its relative lack of potentially confounding exposures. Several aspects of the exposure-response relationship were examined. Different model specifications were used permitting nonlinear dependence on cumulative exposure to be considered, and the possibility of a nonlinear dose-rate effect was also investigated. All models evaluated fit the data comparably well. The linear (additive) relative rate model was selected as the basis for the risk assessment. It was among the better-fitting models and was also preferred on biological grounds because linear low-dose extrapolation is the default assumption for carcinogenesis. There was some suggestion of a negative dose-rate effect, (greater than proportional excess risk at low exposures and less than proportional risk at high exposures but still a monotic relationship) but the effect was small. Although lacking statistical power, the analyses examining thresholds were consistent with no threshold on exposure intensity. Some misclassification on exposure in relation to race appeared to be present, but models with and without the exposure-race interaction produced a clear exposure response. Taken together, the analyses constitute a robust assessment of the risk of chromium carcinogenicity.

The excess lifetime (45 years) risk for lung cancer mortality from exposure to Cr(VI) was estimated to be 255 per thousand workers at the previous OSHA PEL of 52 µg/m^3 based on the exposure-response estimate for all men in the Baltimore cohort. At the previous NIOSH REL of 1 µg/m^3 for Cr(VI) compounds, the excess lifetime risk was estimated to be 6 lung cancer deaths per 1,000 workers and at the REL of 0.2 µg/m^3 the excess lifetime risk is approximately 1 lung cancer death per 1,000 workers.

The data analyzed were from the Baltimore, Maryland cohort previously studied by Hayes et al. [1979] and Gibb et al. [2000b]. The cohort comprised 2,357 men first hired from 1950 through 1974 whose vital status was followed through 1992. The racial makeup of the study population was 1,205 white (51%), 848 non-white (40%) and 304 of unknown race (13%). This cohort had a detailed retrospective exposure assessment that was used to estimate individual worker current and cumulative Cr(VI) exposures across time. Approximately 70,000 both area and personal airborne Cr(VI) measurements of typical exposures were collected and analyzed by the employer from 1950 to 1985, when the plant closed. These samples were used to assign, in successive annual periods, average exposure levels to exposure zones that had been defined by the employer. These job title estimated exposures were combined with individual work histories to calculate the Cr(VI) exposure of each member of the cohort.

Smoking information at hire was available from medical records for 91 percent of the population, including packs per day for 70 percent of the cohort. The cohort was largely free of other potentially confounding exposures. The mean duration of employment of workers in the cohort was 3.1 years, while the median duration was only 0.39 of a year.

In this study population of 2,357 workers, 122 lung cancer deaths were documented. This mortality experience was analyzed using Poisson regression methods. Diverse models of exposure-response for Cr(VI) were evaluated by comparing deviances and inspecting cubic splines. The models using cumulative smoking (as a linear spline) fit significantly better in comparison with models using a simple categorical classification (smoking at hire: yes, no, unknown). For this reason, smoking cumulative exposure imputed from cigarette use at hire was included as a predictor in the final models despite absence of detailed smoking histories. Lifetime risks of lung cancer death from exposure to Cr(VI) were estimated using an actuarial calculation that accounted for competing causes of death.

An additive relative rate model was selected which fit the data well and which was readily interpretable for excess lifetime risk calculations:

$$\text{relative rate} = \exp(\hat{a}_0 + \hat{a}_1 \text{Smk1} + \hat{a}_2 \text{Smk2}) \times (1 + \hat{a}_3 X)$$

where Smk1 and Smk2 are the smoking terms (Smk1, number of pack-years up to 30; and Smk2, above 30) and X is the cumulative chromium exposure (lagged 5 years). The model adjusted for age, race, and calendar time by incorporating national U.S. mortality rates into the model. In the final model, the estimated rate ratio (RR) for 1 mg/m³-yr cumulative exposure to Cr(VI) was 2.44 with a 95% confidence interval of 1.54–3.83 ($\Delta[-2 \ln L]$ = 15.1). Addition of a race-chromium interaction term in the preferred linear relative rate model resulted in a further reduction in deviance of 10.6, a highly statistically significant result ($P = 0.001$), and the observed chromium effect for nonwhite workers (RR = 5.31, 95% CI = 2.78–10.1) was larger than for all workers combined. White workers showed only an overall excess, weakly related to measured cumulative exposure. All the well-fitting models examined had strong race-exposure interactions. This interaction was observed whether age, race, and calendar time were adjusted by stratification (internal adjustment) or by using external population rates. No other important interactions were detected.

A working lifetime of 45 years of exposure to Cr(VI) at the previous OSHA PEL of 100 µg/m³ as CrO_3 corresponds to a cumulative exposure of 4.5 mg/m³-yr. The excess lifetime risk for lung cancer mortality from exposure to Cr(VI) at this exposure level was estimated to be 255 per thousand workers (95% CI: 109–416). At the previous NIOSH REL, 45 years of occupational exposure corresponded to a lifetime excess risk of six (95% CI: 3–12) lung cancer deaths per thousand workers.

Based on a categorical analysis, the exposure-race interaction was found to be largely due to an inverse trend in lung cancer mortality among whites: an excess in the range 0.03–0.09 mg/m³-yr of chromium cumulative exposure and a deficit in the range 0.37–1.1 mg/m³-yr. Park et al. [2004] concluded that a biological basis for the chromium-race interaction was unlikely and that more plausible explanations include, but are not limited to, misclassification of smoking status, misclassification of chromium exposures, or chance. It is doubtful that confounding factors play an important role, because it is unlikely that another causal risk factor is strongly and jointly associated with exposure and race. The asbestos exposure that was present was reported to be typical of industry generally at that time. Some asbestos exposure may have been associated with certain chromium process areas wherein workers were not representative of the entire workforce on race. For this to explain a significant amount of the observed lung cancer excess would require relatively high asbestos exposures correlated with Cr(VI) levels for non-white workers. It would not explain the relative deficit of lung cancer observed among white workers with high cumulative Cr(VI) exposures. Furthermore, no mesothelioma deaths were observed, and the observed lung cancer excess would correspond to asbestos exposures at levels seen only in asbestos manufacturing or processing environments.

Exposure misclassification, on the other hand, is plausible given the well-known disparities in exposure by race often observed in occupational settings. In this study, average exposure levels were assigned to exposure zones within which there may have been substantial race-related differences in work assignments and resulting individual exposures. Race-exposure interactions would inevitably follow. If the racial disparity was the result of exposure misclassification, then models without the race-chromium interaction term would provide an unbiased estimate of the exposure-response, although less precisely than if race had been taken into account in the processing of air-sampling results and in the specification of exposure zone averages.

Park and Stayner [2006] examined the possibility of an exposure threshold in the Baltimore cohort by calculating different measures of cumulative exposure in which only concentrations exceeding some specified threshold value were summed over time. The best–fitting models, evaluated with the profile likelihood method, were those with a threshold lower than 1.0 µg/m^3, the lowest threshold tested. The test was limited by statistical power but established upper confidence limits for a threshold consistent with the observed data of 16 µg/m^3 Cr(VI) for models with the exposure-race interaction or 29 µg/m^3 Cr(VI), for models without the exposure-race interaction. Other models using a cumulative exposure metric in which concentration raised to some power, X^a, is summed over time, found that the best fit corresponded to a = 0.8. If saturation of some protective process were taking place, one would expect a > 1.0. However, statistical power limited interpretation as a = 1.0 could not be ruled out. Analyses in which a cumulative exposure threshold was tested found the best-fitting models with thresholds of 0.02 mg/m^3-yr (with exposure-race interaction) or 0.3 mg/m^3-yr Cr(VI)(without exposure-race interaction) but could not rule out no threshold. The retrospective exposure assessment for the Baltimore cohort, although the best available for a chromium-exposed population, has limitations that reduce the certainty of negative findings regarding thresholds. Nevertheless, the best estimate at this time is that there is no concentration threshold for the Cr(VI)-lung cancer effect.

Crump KS [1995] conducted an analysis of a cohort from the older Baltimore plant reported by Hayes et al. [1979]. The cumulative exposure estimates of Braver et al. [1985] were also used in the risk assessment. From a Poisson regression model, the maximum likelihood estimate of ß, the potency parameter (i.e., unit risk), was 7.5×10^{-4} per µg/m^3-yr. Occupational exposure to Cr(VI) for 45 years was estimated to result in 88 excess lung cancer deaths per 1,000 workers exposed at the previous OSHA PEL and 1.8 excess lung cancer deaths per 1,000 workers exposed at the previous NIOSH REL.

Gibb et al. [1986] conducted a quantitative assessment of the Baltimore production workers reported by Hayes et al. [1979], whose exposure was reconstructed by Braver et al. [1985]. This cohort was divided into six subcohorts based on their period of hire and length of employment [Braver et al. 1985]. Gibb et al. [1986] calculated the lifetime respiratory cancer mortality risk estimates for the four subcohorts who were hired before 1960 and had worked in the old facility. The slopes for these subcohorts ranged from $5.1 \times 10^{-3}/\mu g/m^3$ to $2.0 \times 10^{-2}/\mu g/m^3$ with a geometric mean of $9.4 \times 10^{-3}/\mu g/m^3$.

6.3 Painesville Chromate Production Risk Assessments

Crump et al. [2003] calculated estimates of excess lifetime risk of lung cancer death resulting from occupational and environmental exposure to Cr(VI) in a cohort of chromate chemical production workers. The excess lifetime (45 years) risk for lung cancer mortality from occupational exposure to Cr(VI) at $1 \mu g/m^3$ (the previous NIOSH REL) was estimated to be approximately 2 per 1,000 workers for both the relative and additive risk models.

The cohort analyzed was a Painesville, Ohio worker population described by Luippold et al. [2003]. The cohort comprised 493 workers who met the following criteria: first hired from 1940 through 1972, worked for at least 1 year, and did not work in any of the other Cr(VI) facilities owned by the same company, other than the North Carolina plant. The vital status of the cohort was followed through 1997.

All but four members of the cohort were male. Little information was available on the racial makeup of the study population other than that available from death certificates. Information on potential confounders such as smoking histories and other occupational exposures was limited, so this information was not included in the mortality analysis. There were 303 deaths, including 51 lung cancer deaths, reported in the cohort. SMRs were significantly increased for the following: all causes combined, all cancers combined, lung cancer, year of hire before 1960, 20 or more years of exposed employment, and latency of 20 or more years. A trend test showed a strong relationship between lung cancer mortality and cumulative Cr(VI) exposure. Lung cancer mortality was statistically significantly increased for observation groups with cumulative exposures greater than or equal to 1.05 mg/m³-years.

The exposure assessment of the cohort was reported by Proctor et al. [2003]. More than 800 Cr(VI) air-sampling measurements from 21 industrial hygiene surveys were identified. These data were airborne area samples. Airborne Cr(VI) concentration profiles were constructed for 22 areas of the plant for each month from January 1940 through April 1972. Cr(VI) exposure estimates for each worker were reconstructed by correlating their job titles and work areas with the corresponding area exposure levels for each month of their employment. The cumulative exposure and highest average monthly exposure levels were determined for each worker.

K.S. Crump [1995] calculated the risk of Cr(VI) occupational exposure in its analysis of the Mancuso [1975] data. Cr(III) and Cr(VI) data from the Painesville, Ohio plant [Bourne and Yee 1950] were used to justify a conversion factor of 0.4 to calculate Cr(VI) concentrations from the total chromium concentrations presented by Mancuso [1975]. The cumulative exposure of workers to Cr(VI) (µg/m³-yr) was used in the analysis. All of the original exposure categories presented by Mancuso [1975] were used in the analysis, including those that had the greatest cumulative exposure. A sensitivity analysis using different average values was applied to these exposure categories. U.S. vital statistics data from 1956, 1967, and 1971 were used to calculate the expected numbers of lung cancer deaths. Estimates of excess lung cancer deaths at the previous NIOSH REL ranged from 5.8 to 8.9 per 1,000 workers. Estimates of excess lung cancer deaths at the previous OSHA PEL ranged from 246 to 342 per 1,000 workers.

DECOS [1998] used the EPA [1984] environmental risk assessment that was based on the Mancuso [1975] data to calculate the additional lung cancer mortality risk due to occupational Cr(VI) exposure. The EPA estimate that occupational exposure to 8 µg/m³ total dust resulted in an additional lung cancer mortality risk of 1.4×10^{-2} was used to calculate occupational risk. It was assumed that total dust concentrations were similar to inhalable dust concentrations because of the small aerodynamic diameters of the particulates. Additional cancer mortality risks for 40-year occupational exposure to inhalable dust were calculated as 4×10^{-3} for 2 µg/m³ Cr(VI).

The EPA used the data of Mancuso [1975] to calculate a unit risk estimate for Cr(VI). A unit risk estimate is the incremental lifetime cancer risk over the background cancer risk occurring in a hypothetical population in which all individuals are exposed continuously throughout life to a concentration of 1 µg/m³ of the agent in the air that they breathe [EPA 1984]. This unit risk quantifies the risk resulting from environmental exposure to Cr(VI) as an air pollutant. The U.S. EPA calculated, based on the Mancuso [1975] data, a unit risk estimate for Cr(VI) of 1.2×10^{-2} for environmental exposures. If this lifetime unit risk estimate is adjusted to a hypothetical working lifetime of Cr(VI) exposure (8-hour work day, 250 days per year for 45 years), there would be 92.5 predicted additional deaths from lung cancer per 1,000 workers at the previous OSHA PEL of 52 µg/m³ and 1.8 predicted additional deaths from lung cancer per 1,000 workers at the previous NIOSH REL of 1 µg/m³ [Crump 1995].

6.4 Other Cancer Risk Assessments

The International Chromium Development Association (ICDA) [1997] used the overall SMR for lung cancer from 10 Cr(VI) studies to assess the risk of occupational exposure to various levels of Cr(VI) exposure. The 10 studies evaluated were those selected by Steenland et al. [1996] as the largest and best-designed studies of workers in the chromium production, chromate paint production, and chromate plating industries. It was assumed that the mean length of employment of all workers was 15 years. Although this assumption may be appropriate for some of the cohorts, for others it is not: the mean duration of employment for the Painesville cohort was less than 10 years, and for the Baltimore cohort it was less than 4 years. Occupational exposures to Cr(VI) were assumed to be 500 µg/m³, 1,000 µg/m³, or 2,000 µg/m³ TWA. These are very unlikely Cr(VI) exposure levels. The mean exposure concentrations in the Painesville cohort were less than 100 µg/m³ after 1942, and in the Baltimore cohort the mean exposure concentration was 45 µg/m³. For these different exposure levels, three different assumptions were tested: (1) the excess SMR was due to only Cr(VI) exposure, (2) Cr(VI) exposure was confounded by smoking or other occupational exposures so that the baseline SMR should be 130, or (3) confounders set the baseline SMR to 160. The investigators did not adjust for the likely presence of a healthy worker effect in these SMR analyses. A baseline SMR of 80 or 90 would have been appropriate based on other industrial cohorts and would have addressed smoking differences between industrial worker populations and national reference populations [Park et al. 1991]. The reference used for expected deaths was the 1981 life-table for males in England and Wales. The lung cancer mortality risk estimates ranged from 5 to 28 per 1,000 at exposure to 50 µg/m³ Cr(VI), to 0.1 to 0.6 per 1,000 at exposure to 1 µg/m³ Cr(VI). The assumptions made and methods used in this risk assessment make it a weaker analysis than those in which worker exposure data at a particular plant are correlated with their incidence of lung cancer. The excess lung cancer deaths may have been underestimated by at least a factor of 10, given the assumptions used on duration (factor of 1.5–2.0), exposure level (factor of 10–20), and healthy worker bias (factor of 1.1–1.2).

6.5 Summary

The data sets of the Painesville, Ohio and Baltimore, Maryland chromate production workers provide the bases for the quantitative risk assessments of excess lung cancer deaths due to occupational Cr(VI) exposure. In 1975, Mancuso presented the first data set of the Painesville, Ohio workers, which was used for quantitative risk analysis. Its deficiencies included very limited exposure data, information on total chromium only, and no reporting of the expected number of deaths from lung cancer. Proctor et al. [2003] presented more than 800 airborne Cr(VI) measurements from 23 newly identified surveys conducted from 1943 through 1971 at the Painesville plant. These data and the mortality study of Luippold et al. [2003] provided the basis for an improved lung cancer risk assessment of the Painesville workers.

In 1979, Hayes presented the first data of the Baltimore, Maryland production facility workers, which were later used for quantitative risk assessment. In 2000, Gibb and coworkers provided additional exposure data for an improved cancer risk assessment of the Baltimore workers [Gibb et al. 2000b]. NIOSH selected the Gibb et al. [2000b] cohort for quantitative risk analysis [Park et al. 2004] rather than the Painesville cohort because of its greater number of lung cancer deaths, better smoking histories, and a more comprehensive retrospective exposure archive [NIOSH 2005a].

In spite of the different data sets analyzed and the use of different assumptions, models, and calculations, these risk assessments have estimates of excess risk that are within an order of magnitude of each other (see Tables 6–1, 6–2). All of these risk assessments indicate considerable excess risk of lung cancer death to workers exposed to Cr(VI) at the previous OSHA PEL and previous NIOSH REL. The risk assessments of Crump et al. [2003] and Park et al. [2004] analyzed the most complete data sets available on occupational exposure to Cr(VI). These risk assessments estimated excess risks of lung cancer death of 2 per 1,000 workers [Crump et al. 2003] and 6 per 1,000 workers [Park et al. 2004], at a working lifetime exposure to 1 µg/m³. Park et al. [2004] estimated an excess risk of lung cancer death of approximately 1 per 1,000 workers at a steady 45-year workplace exposure to 0.2 µg/m³ Cr(VI).

Park and Stayner [2006] evaluated the possibility of a threshold concentration for lung cancer in the Baltimore cohort. Although a threshold could not be ruled out because of the limitations of the analysis, the best estimate at this time is that there is no concentration threshold for the Cr(VI)-lung cancer effect.

Table 6-1. Risk Assessments of the Painesville Cohort
Estimated additional deaths from lung cancer per 1,000 workers

Cr(VI) exposure µg/m³*	EPA [1984]	Crump [1995]§	Crump et al. [2003]
0.25	0.44	1.4–2.2	
0.5		2.9–4.4	
1.0†	1.8	5.8–8.9	2.1 (1.3–2.9); 2.2 (1.4–3.0)¶
2.5	4.4	14.0–22.0	
5.0‡	8.8	28.0–43.0	
52.0	91.5	246–342	

*Assumes steady working lifetime exposure
†Previous NIOSH REL
‡OSHA PEL
§Range results from different treatments of high-exposure groups
¶Result (95% confidence interval) for relative risk and additive risk models, respectively.

Table 6-2. Risk Assessments of the Baltimore Cohorts
Estimated additional deaths from lung cancer per 1,000 workers

Cr(VI) exposure µg/m³*	Gibb et al. [1986]	KS Crump [1995]	Park et al. [2004] linear model	Park et al. [2004] log-linear model
0.25	0.34	0.45	1.5	—
0.5	—	0.90	3 (1–6)§	3 (1–4)
1.0†	1.4	1.8	6 (3–12)	5 (3–8)
2.5	3.4	4.5	16 (6–30)	14 (7–20)
5.0‡	6.8	9.0	31 (12–59)	28 (13–43)
52.0	70.2	88.0	255 (109–416)	281 (96–516)

*Assumes steady working lifetime exposure
†Previous NIOSH REL
‡OSHA PEL
§95% confidence interval

7 Recommendations for an Exposure Limit

NIOSH is mandated under the authority of the Occupational Safety and Health Act of 1970 (Public Law 91-596) to develop and recommend criteria for identifying and controlling workplace hazards that may result in occupational illness or injury. NIOSH evaluated the available literature on hexavalent chromium (Cr[VI]) compounds, including quantitative risk assessment, epidemiologic, toxicologic, and industrial hygiene studies to develop recommendations for occupational exposure to Cr(VI) compounds. This chapter summarizes the information relevant to the NIOSH recommended exposure limit (REL) for Cr(VI) compounds and the scientific data used to derive and support the revised REL. More detailed information about the studies summarized here is available in the respective document chapters.

7.1 The NIOSH REL for Cr(VI) Compounds

NIOSH recommends that airborne exposure to all Cr(VI) compounds be limited to a concentration of 0.2 µg Cr(VI)/m^3 for an 8-hr TWA exposure during a 40-hr workweek. The use of NIOSH Method 7605 (or validated equivalents) is recommended for Cr(VI) determination. The REL represents the upper limit of exposure for each worker during each work shift. Because of the residual risk of lung cancer at the REL, NIOSH further recommends that all reasonable efforts be made to reduce exposures to Cr(VI) compounds below the REL. The available scientific evidence supports the inclusion of all Cr(VI) compounds into this recommendation. The REL is intended to reduce workers' risk of lung cancer associated with occupational exposure to Cr(VI) compounds over a working lifetime. It is expected that reducing airborne workplace exposures will also reduce the nonmalignant respiratory effects of Cr(VI) compounds, including irritated, ulcerated, or perforated nasal septa and other potential adverse health effects. Additional controls are needed or administrative actions should be taken to reduce 8-hr TWA exposure to Cr(VI) compounds when the results of the exposure monitoring plan do not produce a high degree of confidence that a high percentage of daily 8-hr TWA exposures are below the REL.

In addition to limiting airborne concentrations of Cr(VI) compounds, NIOSH recommends that dermal exposure to Cr(VI) be prevented in the workplace to reduce the risk of adverse dermal health effects, including irritation, ulcers, skin sensitization, and allergic contact dermatitis.

7.2 Basis for NIOSH Standards

In the 1973 *Criteria for a Recommended Standard: Occupational Exposure to Chromic Acid*, NIOSH recommended that the federal standard for chromic acid, 0.1 mg/m^3 as a 15- minute ceiling concentration, be retained because of reports of nasal ulceration occurring at concentrations only slightly above this concentration [NIOSH 1973a]. In addition, NIOSH recommended supplementing the ceiling concentration with a TWA of 0.05 mg/m^3 for an 8-hour workday to protect against possible chronic effects, including lung cancer and liver damage. The association of these chronic effects with chromic acid exposure was not proven at that time, but the possibility of a correlation could not be rejected [NIOSH 1973a].

In the 1975 *Criteria for a Recommended Standard: Occupational Exposure to Chromium(VI)*, NIOSH supported two distinct recommended standards for Cr(VI) compounds [NIOSH 1975a]. Some Cr(VI) compounds were considered noncarcinogenic at that time, including the chromates and bichromates of hydrogen, lithium, sodium, potassium, rubidium, cesium, and ammonium, and chromic acid anhydride. These Cr(VI) compounds were relatively soluble in water. It was recommended that a 10-hr TWA limit of 25 µg Cr(VI)/m^3 and a 15-minute ceiling limit of 50 µg Cr(VI)/m^3 be applied to these Cr(VI) compounds.

All other Cr(VI) compounds were considered carcinogenic [NIOSH 1975a]. These Cr(VI) compounds were relatively insoluble in water. At that time, NIOSH had a carcinogen policy that called for "no detectable exposure levels for proven carcinogenic substances" [Fairchild 1976]. Thus the basis for the REL for carcinogenic Cr(VI) compounds, 1 µg Cr(VI)/m^3 TWA, was the quantitative limitation of the analytical method available for measuring workplace exposures to Cr(VI) at that time.

NIOSH revised its policy on Cr(VI) compounds in its 1988 *Testimony to OSHA on the Proposed Rule on Air Contaminants* [NIOSH 1988b]. NIOSH testified that while insoluble Cr(VI) compounds had previously been demonstrated to be carcinogenic, there was now sufficient evidence that soluble Cr(VI) compounds were also carcinogenic. Human studies cited in support of this position included Blair and Mason [1980], Franchini et al. [1983], Royle [1975a,b], Silverstein et al. [1981], Sorahan et al. [1987], and Waterhouse [1975]. In addition, the animal studies of Glaser et al. [1986] and Steinhoff et al. [1986] were cited as demonstrating that lifespan exposure of rats to soluble chromates could induce statistically significant excess cancer rates. NIOSH recommended that all Cr(VI) compounds, whether soluble or insoluble, be classified as potential occupational carcinogens. NIOSH recommended that OSHA adopt the most protective of the available standards, the NIOSH RELs. Subsequently, based on this testimony to OSHA, the REL of 1 µg Cr(VI)/m^3 TWA was adopted by NIOSH for all Cr(VI) compounds.

NIOSH reaffirmed its policy that all Cr(VI) compounds be classified as occupational carcinogens in the *NIOSH Comments on the OSHA Request for Information on Occupational Exposure to Hexavalent Chromium* and the *NIOSH Testimony on the OSHA Proposed Rule on Occupational Exposure to Hexavalent Chromium* [NIOSH 2002; 2005a].

This criteria document describes the most recent NIOSH scientific evaluation of occupational exposure to Cr(VI) compounds, including the justification for a revised REL derived using current quantitative risk assessment methodology on human health effects data. Derivation of the REL follows the criteria established by NIOSH in 1995 in which RELs, including those for carcinogens, would be based on risk evaluations using human or animal health effects data, and on an assessment of what levels can be feasibly achieved by engineering controls and measured by analytical techniques [NIOSH 1995b].

7.3 Evidence for the Carcinogenicity of Cr(VI) Compounds

Hexavalent chromium is a well-established occupational carcinogen associated with lung cancer and nasal and sinus cancer [ATSDR 2012; EPA 1998; IARC 1990, 2012; Straif et al. 2009]. NTP identified Cr(VI) compounds as carcinogens in its first report on carcinogens in 1980 [NTP 2011]. Toxicologic studies, epidemiologic studies, and lung cancer meta-analyses provide evidence for the carcinogenicity of Cr(VI) compounds.

7.3.1 Epidemiologic Lung Cancer Studies

In 1989, the IARC critically evaluated the published epidemiologic studies of chromium compounds including Cr(VI) and concluded that "there is sufficient evidence in humans for the carcinogenicity of chromium[VI] compounds as encountered in the chromate production, chromate pigment production and chromium plating industries" (i.e., IARC category "Group 1" carcinogen) [IARC 1990]. Results from two recent lung cancer mortality studies of chromate production workers support this evaluation [Gibb et al. 2000b; Luippold et al. 2003]. In 2009, an IARC Working Group reviewed and reaffirmed Cr(VI) compounds as Group 1 carcinogens (lung) [Straif et al. 2009; IARC 2012].

Gibb et al. [2000b] conducted a retrospective analysis of lung cancer mortality in a cohort of Maryland chromate production workers. The cohort of 2,357 male workers first employed from 1950 through 1974 was followed until 1992. Workers with short-term employment (i.e., < 90 days) were included in the study group to increase the size of the low exposure group. The mean length of employment was 3.1 years. A detailed retrospective assessment of Cr(VI) exposure based on more than 70,000 personal and area samples (short-term and full-shift) and information about most workers' smoking habits at hire was available.

Lung cancer standardized mortality ratios increased with increasing cumulative exposure (i.e., mg CrO_3/m^3-years, with 5-year exposure lag)—from 0.96 in the lowest quartile to 1.57 (95% CI 1.07–2.20) and 2.24 (95% CI 1.60–3.03) in the two highest quartiles. The number of expected lung cancer deaths was based on age-, race-, and calendar year-specific rates for Maryland. Proportional hazards models that controlled for the effects of smoking predicted increasing lung cancer risk with increasing Cr(VI) cumulative exposure (relative risks: 1.83 for second exposure quartile, 2.48 for third, and 3.32 for fourth, compared with first quartile of cumulative exposure; confidence intervals not reported; 5-year exposure lag).

Luippold et al. [2003] conducted a retrospective cohort study of lung cancer mortality in 493 chromate production workers employed for at least 1 year from 1940 through 1972 in a Painesville, Ohio plant. Their mortality was followed from 1941 to the end of 1997 and compared with United States and Ohio rates. The effects of smoking could not be assessed because of insufficient data. More than 800 area samples of airborne Cr(VI) from 21 industrial hygiene surveys were available for formation of a job-exposure matrix [Proctor et al. 2003]. Cumulative Cr(VI) exposure was divided into five categories: 0.00–0.19, 0.20–0.48, 0.49–1.04, 1.05–2.69, and 2.70–23.0 mg/m³-years [Luippold et al. 2003]. Person-years in each category ranged from 2,369 to 3,220, and the number of deaths from trachea, bronchus, or lung cancer ranged from 3 in the lowest exposure category to 20 in the highest (n = 51). The SMRs were statistically significant in the two highest cumulative exposure categories (3.65 [95% CI 2.08–5.92] and 4.63 [2.83–7.16], respectively). SMRs were also significantly increased for year of hire before 1960, ≥ 20 years of employment, and ≥ 20 years since first exposure. The tests for trend across increasing categories of cumulative exposure, year of hire, and duration of employment were statistically significant ($P \leq 0.005$). A test for departure of the data from linearity was not statistically significant (χ^2 goodness of fit of linear model; $P = 0.23$).

7.3.2 Lung Cancer Meta-Analyses

Meta-analyses of epidemiologic studies have been conducted to investigate cancer risk in chromium-exposed workers. Most of these studies also provide support for the classification of Cr(VI) compounds as occupational lung carcinogens.

Sjögren et al. [1994] reported a meta-analysis of five lung cancer studies of Canadian and European welders exposed to stainless steel welding fumes. The meta-analysis found an estimated relative risk of 1.94 (95% CI 1.28–2.93) and accounted for the effects of smoking and asbestos exposure.

Steenland et al. [1996] reported overall relative risks for specific occupational lung carcinogens identified by IARC, including chromium. Ten epidemiologic studies were selected by the authors as the largest and best-designed studies of chromium production workers, chromate pigment production workers, and chromium platers. The summary relative risk for the 10 studies was 2.78 (95% confidence interval 2.47–3.52; random effects model), which was the second-highest relative risk among the eight carcinogens summarized.

Cole and Rodu [2005] conducted meta-analyses of epidemiologic studies published in 1950 or later to test for an association of chromium exposure with all causes of death, and death from malignant diseases (i.e., all cancers combined, lung cancer, stomach cancer, cancer of the central nervous system [CNS], kidney cancer, prostate gland cancer, leukemia, Hodgkin's disease, and other lymphohematopoietic cancers). Available papers (n = 114) were evaluated independently by both authors on eight criteria that addressed study quality. In addition, papers with data on lung cancer were assessed for control of cigarette smoking effects, and papers with data on stomach cancer were assessed for economic status. Forty-nine epidemiologic studies based on 84 papers published were used in the meta-analyses. The number of studies in each meta-analysis ranged from 9 for Hodgkin's disease to 47 for lung cancer. Association was measured by an author-defined "SMR" that included odds ratios, proportionate mortality ratios, and most often, standardized mortality ratios. Mortality risks were not significantly increased for most causes of death. However, SMRs were significantly increased in all lung cancer meta-analyses (smoking controlled: 26 studies; 1,325 deaths; SMR = 118; 95% CI 112–125) (smoking not controlled: 21 studies; 1,129 deaths; SMR = 181; 95% CI 171–192) (lung cancer—all: 47 studies; 2,454 deaths; SMR = 141; 95% CI 135–147). Stomach cancer mortality risk was significantly increased only in meta-analyses of studies that did not control for effects of economic status (economic status not controlled: 18 studies; 324 deaths; SMR = 137; 95% 123–153). The authors stated that statistically significant SMRs for "all cancer" mortality were mainly due to lung cancer (all cancer: 40 studies; 6,011 deaths; SMR = 112; 95% CI 109–115). Many of the studies contributing to the meta-analyses did not address bias from the healthy worker effect, and thus the results are likely underestimates of the cancer mortality risks. Other limitations of these meta-analyses include lack of (1) exposure characterization of populations, such as the route of exposure (i.e., airborne versus ingestion); and (2) detail of criteria used to exclude studies based on "no or little chrome exposure" or "no usable data."

7.3.3 Animal Experimental Studies

Cr(VI) compounds have been tested in animals using many different experimental conditions and exposure routes. Although experimental conditions are often different from occupational exposures, these studies provide additional data to assess the carcinogenicity of the test compounds. Chronic inhalation studies provide the best data for extrapolation to occupational exposure; few have been conducted using Cr(VI) compounds. However, the body of animal studies support the classification of Cr(VI) compounds as occupational carcinogens.

The few chronic inhalation studies available demonstrate the carcinogenic effects of Cr(VI) compounds in mice and rats [Adachi et al. 1986, 1987; Glaser et al. 1986]. Female mice exposed to 1.8 mg/m^3 chromic acid mist (2 hours per day, 2 days per week for up to 12 months) developed a significant number of nasal papillomas compared with control animals [Adachi 1987]. Female mice exposed to a higher dose of chromic acid mist, 3.6 mg/m^3 (30 minutes per day, 2 days per week for up to 12 months) developed an increased, but not statistically significant, number of lung adenomas [Adachi et al. 1986]. Glaser et al. [1986] reported a statistically significant number of lung tumors in male rats exposed for 18 months to 100 μg/m^3 sodium dichromate; no tumors were reported at lower dose levels.

Animal studies conducted using other routes of administration have also produced adverse health effects with some Cr(VI) compounds. Zinc chromate and calcium chromate produced a statistically significant ($P < 0.05$) number of bronchial carcinomas when administered to rats via an intrabronchial pellet implantation system [Levy et al. 1986]. Cr(VI) compounds with a range of solubilities were tested using this system. Although some soluble Cr(VI) compounds did produce bronchial carcinomas, these results were not statistically significant. Some lead chromate compounds produced bronchial squamous carcinomas that, although not statistically significant, may be biologically significant because of the absence of this cancer in control rats.

Steinhoff et al. [1986] administered the same total dose of sodium dichromate either once per week or five times per week to male and female rats via intratracheal instillation. No increased incidence of lung tumors was observed in animals dosed five times weekly. However, in animals dosed once per week, a statistically significant tumor incidence was reported in the 1.25 mg/kg exposure group. This study demonstrates a dose-rate effect within the constraints of the experimental design. It suggests that limiting exposure to high Cr(VI) concentrations may be important in reducing carcinogenicity. However, quantitative extrapolation of these animal data to the human exposure scenario is difficult.

Animal studies conducted using nonrespiratory routes of administration have also produced injection-site tumors with some Cr(VI) compounds [Hueper 1961; Furst 1976]. These studies provide another data set for hazard identification.

IARC [2012] concluded "there is sufficient evidence in experimental animals for the carcinogenicity of chromium (VI) compounds".

7.4 Basis for the NIOSH REL

The primary basis for the NIOSH REL is the results of the Park et al. [2004] quantitative risk assessment of lung cancer deaths of Maryland chromate production workers conducted on the data of Gibb et al. [2000b]. NIOSH determined that this was the best Cr(VI) data set available for analysis because of its extensive exposure assessment and smoking information, strong statistical power, and its relative lack of potentially confounding exposures [NIOSH 2005a]. The results of the NIOSH risk assessment are supported by other quantitative Cr(VI) risk assessments (see Chapter 6).

NIOSH selected the revised REL at an excess risk of lung cancer of approximately 1 per 1,000 workers based on the results of Park et al. [2004]. Table 7–1 presents the range of risk levels of lung cancer from 1 per 500 to 1 per 100,000 for workers exposed to Cr(VI). Cancer risks greater than 1 per 1,000 are considered significant and worthy of intervention by OSHA. This level of risk is consistent with those for other carcinogens in recent OSHA rules [71 Fed. Reg. 10099 (2006)]. NIOSH has used this risk level in a variety of circumstances, including citing this level as appropriate for developing authoritative recommendations in criteria documents and peer-reviewed risk assessments [NIOSH 1995a, 2006; Rice et al. 2001; Park et al. 2002; Stayner et al. 2000; Dankovic et al. 2007].

Additional considerations in the derivation of the REL include analytical feasibility and the ability to control exposure concentrations to the REL in the workplace. The REL for Cr(VI) compounds is intended to reduce workers' risk of lung cancer over a 45-year working lifetime. Although the quantitative analysis is based on lung cancer mortality data, it is expected that reducing airborne workplace exposures will also reduce the nonmalignant respiratory effects of Cr(VI) compounds, including irritated, ulcerated, or perforated nasal septa, and other potential adverse health effects.

Table 7–1. Cr(VI) exposure associated with various levels of excess risk of lung cancer after a 45-year working lifetime

Lifetime added risk* for a 45-year working lifetime exposure	Cr(VI) exposure $\mu g/m^3$ Cr	Cr(VI) exposure $\mu g/m^3$ CrO_3
1 in 500	0.32	0.64
1 in 1,000	0.16	0.32
1 in 2,000	0.08	0.16
1 in 5,000	0.032	0.064
1 in 10,000	0.016	0.032
1 in 100,000	0.0016	0.0032

*Risk estimates from Park et al. [2004]

The available scientific evidence supports the inclusion of workers exposed to all Cr(VI) compounds into this recommendation. All Cr(VI) compounds studied have demonstrated their carcinogenic potential in animal, in vitro, or human studies [NIOSH 1988b; 2002; 2005a,b]. Molecular toxicology studies provide additional support for classifying Cr(VI) compounds as occupational carcinogens.

At this time, there is insufficient data to quantify a different REL for each specific Cr(VI) compound [NIOSH 2005a,b]. Although there are inadequate epidemiologic data to quantify the risk of human exposure to insoluble Cr(VI) compounds, the results of animal studies indicate that this risk is likely as great as, if not greater than, exposure to soluble Cr(VI) compounds [Levy et al. 1986]. Because of the similar mechanisms of action of soluble and insoluble Cr(VI) compounds, and the quantitative risk assessments demonstrating significant risk of lung cancer death resulting from occupational lifetime exposure to soluble Cr(VI) compounds, NIOSH recommends that the REL apply to all Cr(VI) compounds.

At this time, there are inadequate data to conduct a quantitative risk assessment for workers exposed to Cr(VI), other than chromate production workers. However, epidemiologic studies demonstrate that the health effects of airborne exposure to Cr(VI) are similar across workplaces and industries (see Chapter 4). Therefore, the results of the NIOSH quantitative risk assessment conducted on chromate production workers [Park et al. 2004] are being used as the basis of the REL for workplace exposures to all Cr(VI) compounds.

7.4.1 Park et al. [2004] Risk Assessment

NIOSH calculated estimates of excess lifetime risk of lung cancer death resulting from occupational exposure to water-soluble chromium-containing mists and dusts in a cohort of Baltimore, MD chromate chemical production workers [Park et al. 2004]. This cohort, originally studied by Gibb et al. [2000b], comprised 2,357 men first hired from 1950 through 1974, whose vital status was followed through 1992. The mean duration of employment of workers in the cohort was 3.1 years, and the median duration was 0.39 year.

This cohort had a detailed retrospective exposure assessment of approximately 70,000 measurements, which was used to estimate individual worker current and cumulative Cr(VI) exposures across time. Smoking information at hire was available from medical records for 91% of the population, including packs per day for 70 percent of the cohort. In this study population of 2,357 workers, 122 lung cancer deaths were documented.

The excess working lifetime (45 years) risk estimates of lung cancer death associated with occupational exposure to water-soluble Cr(VI) compounds using the linear risk model are 255 (95% CI: 109–416) per 1,000 workers at 52 µg Cr(VI)/m^3, 6 (95% CI: 3–12) per 1,000 workers at 1 µg Cr(VI)/m^3, and approximately 1 per 1,000 workers at 0.2 µg Cr(VI)/m^3.

7.4.2 Crump et al. [2003] Risk Assessment

Crump et al. [2003] analyzed data from the Painesville, Ohio chromate production worker cohort described by Luippold et al. [2003]. The cohort comprised 493 workers who met the following criteria: first hired from 1940 through 1972, worked for at least 1 year, and did not work in any of the other Cr(VI) facilities owned by the same company other than the North Carolina plant. The vital status of the cohort was followed through 1997.

Information on potential confounders (e.g., smoking) and other occupational exposures was limited and not included in the mortality analysis. There were 303 deaths reported, including 51 lung cancer deaths. SMRs were significantly increased for the following: all causes combined, all cancers combined, lung cancer, year of hire before 1960, 20 or more years of exposed employment, and latency of 20 or more years. A trend test showed a strong relationship between lung cancer mortality and cumulative Cr(VI) exposure. Lung cancer mortality was increased for cumulative exposures greater than or equal to 1.05 mg/m^3-years.

The estimated lifetime additional risk of lung cancer mortality associated with 45 years of occupational exposure to water-soluble Cr(VI) compounds at 1 µg/m^3 was approximately 2 per 1,000 (0.00205 [90% CI: 0.00134, 0.00291] for the relative risk model and 0.00216 [90% CI: 0.00143, 0.00302] for the additive risk model), assuming a linear dose response for cumulative exposure with a 5-year lag.

7.4.3 Risk Assessment Summary

Quantitative risk assessments of the Baltimore, Maryland and Painesville, Ohio chromate production workers consistently demonstrate significant risk of lung cancer mortality to workers exposed to Cr(VI) at the previous NIOSH REL of 1 µg Cr(VI)/m^3. These results justify lowering the NIOSH REL to decrease the risk of lung cancer in workers exposed to Cr(VI). NIOSH used the results of the risk assessment of Park et al. [2004] as the basis of the current REL because this assessment analyzes the most extensive database of workplace exposure measurements, including smoking data on most workers.

7.5 Applicability of the REL to All Cr(VI) Compounds

NIOSH recommends that the REL of 0.2 µg Cr(VI)/m^3 be applied to all Cr(VI) compounds. Currently, there are inadequate data to exclude any single Cr(VI) compound from this recommendation. IARC [2012] concluded that "there is sufficient evidence in humans for the carcinogenicity of chromium (VI) compounds".

Epidemiologic studies were often unable to identify the specific Cr(VI) compound responsible for the excess risk of cancer. However, these studies have documented the carcinogenic risk of occupational exposure to soluble Cr(VI). Gibb et al. [2000b] and Luippold et al. [2003] reported the health effects of chromate production workers with sodium dichromate being their primary Cr(VI) exposure. These studies, and the risk assessments conducted on their data, demonstrate the carcinogenic effects of this soluble Cr(VI) compound. The NIOSH risk assessment on which the REL is based evaluated the risk of exposure to sodium dichromate [Park et al. 2004].

Although there are inadequate epidemiologic data to quantify the cancer risk of human exposure to insoluble Cr(VI) compounds, the results of animal studies indicate that this risk is likely as great, if not greater than, exposure to soluble Cr(VI) compounds [Levy et al. 1986]. The carcinogenicity of insoluble Cr(VI) compounds has been demonstrated in animal and human studies [NIOSH 1988b]. Animal studies have demonstrated the carcinogenic potential of soluble and insoluble Cr(VI) compounds [NIOSH 1988b, 2002, 2005a; ATSDR 2012]. IARC [2012] concluded that "there is sufficient evidence in experimental animals for the carcinogencity of chromium (VI) compounds". Based on the current scientific evidence, NIOSH recommends including all Cr(VI) compounds in the revised REL. There are inadequate data to exclude any single Cr(VI) compound from this recommendation.

7.6 Analytical Feasibility of the REL

Several validated methods can quantify airborne exposures to Cr(VI) in workplace air. NIOSH Method 7605, OSHA Method ID-215, and international consensus standard analytical methods can quantitatively assess worker exposure to Cr(VI) at the REL of 0.2 µg Cr(VI)/m^3. The LOD for NIOSH Method 7605 is 0.02 µg per sample [NIOSH 2003b]. Sampling considerations to ensure accurate workplace Cr(VI) measurements are discussed in Chapter 3.

7.7 Controlling Workplace Exposures

Elimination of and substitution for Cr(VI) compounds, and the use of good work practices and engineering controls, should be the highest priorities for controlling Cr(VI) exposures. Control techniques such as source enclosure (i.e., isolating the generation source from the worker) and LEV systems are the preferred methods for preventing worker exposure to airborne hazards. OSHA determined that the primary engineering control measures most likely to be effective in reducing worker exposure to airborne Cr(VI) are LEV, process enclosure, process modification, and improving general dilution ventilation [71 Fed. Reg. 10099 (2006)]. Section 8.3 provides additional information and recommendations for exposure control measures.

The NIOSH REL is a health-based recommendation. Additional considerations include analytical feasibility and the achievability of engineering controls. Based on a qualitative assessment of workplace exposure data, NIOSH acknowledges Cr(VI) exposures below the REL can be achieved in some workplaces using existing technologies, but exposures are more difficult to control in other workplaces. Some operations, including hard chromium electroplating, chromate-paint spray application, atomized-alloy spray-coating, and welding may have difficulty in consistently achieving exposures at or below the REL by means of work practices and engineering controls (see Table 2–7) [Blade et al. 2007]. The extensive industry analysis of workplace exposures conducted for the OSHA rule-making process supports the NIOSH assessment that the REL is achievable in some workplaces but difficult to achieve in others (see Table 2–8) [71 Fed. Reg. 10099 (2006)]. The Cr(VI) REL is intended to promote the proper use of existing control technologies and encourage the development of new control technologies where needed to control workplace Cr(VI) exposures. The consistent and proper use of control technologies will continue to reduce workplace Cr(VI) exposures.

NIOSH acknowledges that the frequent use of PPE, including respirators, may be required by some workers in environments where airborne Cr(VI) concentrations cannot be controlled below the REL in spite of implementing all other possible measures in the hierarchy of controls. The frequent use of PPE may be required during job tasks for which (1) routinely high airborne concentrations of Cr(VI) exist, (2) the airborne concentration of Cr(VI) is unknown or unpredictable, and (3) job tasks are associated with highly variable airborne concentrations because of environmental conditions or the manner in which the job task is performed.

7.8 Preventing Dermal Exposure

NIOSH recommends that dermal exposure to Cr(VI) be prevented by elimination or substitution of Cr(VI) compounds. When this is not possible, appropriate sanitation and hygiene procedures and appropriate PPE should be used (see Chapter 8). Preventing dermal exposure is important to reduce the risk of adverse dermal health effects, including dermal irritation, ulcers, skin sensitization, and allergic contact dermatitis. The prevention of dermal exposure to Cr(VI) compounds is critical in preventing skin disorders related to Cr(VI).

7.9 Summary

NIOSH determined that the data of Gibb et al. [2000b] is the most comprehensive data set available for assessing the health risk of occupational exposure to Cr(VI), including an extensive exposure assessment database and smoking information on workers. The revised REL is a health-based recommendation derived from the results of the NIOSH quantitative risk assessment conducted on these human health effects data [Park et al. 2004]. Other considerations include analytical feasibility and the achievability of engineering controls.

NIOSH recommends a REL of 0.2 µg Cr(VI)/m^3 for an 8-hr TWA exposure within a 40-hr workweek, for all airborne Cr(VI) compounds. The REL is intended to reduce workers' risk of lung cancer over a 45-year working lifetime. The excess risk of lung cancer death at the revised REL is approximately 1 per 1,000 workers. NIOSH has used this risk level in other authoritative recommendations in criteria documents and peer-reviewed risk assessments. Results from epidemiologic and toxicologic studies provide the scientific evidence to classify all Cr(VI) compounds as occupational carcinogens and support the recommendation of having one REL for all Cr(VI) compounds [NIOSH 1988b, 2002, 2005a,b].

Exposure to Cr(VI) compounds should be eliminated from the workplace where possible because of the carcinogenic potential of these compounds. Where possible, less-toxic compounds should be substituted for Cr(VI) compounds. Where elimination or substitution of Cr(VI) compounds is not possible, attempts should be made to control workplace exposures below the REL. Compliance with the REL for Cr(VI) compounds is currently achievable in some industries and for some job tasks. It may be difficult to achieve the REL during certain job tasks including welding, electroplating, spray painting, and atomized-alloy spray-coating operations. Where airborne exposures to Cr(VI) cannot be reduced to the REL through using state-of-the-art engineering controls and work practices, the use of respiratory protection will be needed.

The REL may not be sufficiently protective to prevent all occurrences of lung cancer and other adverse health effects among workers exposed for a working lifetime. NIOSH therefore recommends that worker exposures be maintained as far below the REL as achievable during each work shift. NIOSH also recommends that a comprehensive safety and health program be implemented that includes worker education and training, exposure monitoring, and medical monitoring.

In addition to controlling airborne exposures at the REL, NIOSH recommends that dermal exposures to Cr(VI) compounds be prevented to reduce the risk of adverse dermal health effects, including dermal irritation, ulcers, skin sensitization, and allergic contact dermatitis.

8 Risk Management

NIOSH recommends the following guidelines to minimize worker exposure to hexavalent chromium (Cr[VI]) compounds. Adherence to these recommendations should decrease the risk of lung cancer in workers exposed to Cr(VI) compounds. It is expected that reducing airborne workplace exposures will also reduce the nonmalignant respiratory effects of Cr(VI) compounds, including irritated, ulcerated, or perforated nasal septa and other potential adverse health effects. Although workplaces in which workers are exposed to Cr(VI) levels above the recommended exposure limit (REL) warrant particular concern and attention, all workplaces should attempt to decrease worker exposures to Cr(VI) compounds to the lowest level that is reasonably achievable in order to minimize adverse health effects, including lung cancer, in workers. The following recommendations should be incorporated into a comprehensive safety and health plan in each workplace in which workers manufacture, use, handle, or dispose of Cr(VI) compounds, or perform any other activity that involves exposure to Cr(VI) compounds.

In 2006, OSHA amended its standard for occupational exposure to Cr(VI) compounds [71 Fed. Reg. 10099 (2006)]. The final standard separately regulates general industry, construction, and shipyards in order to tailor requirements to the unique circumstances found in each of these industry sectors. For a full list and explanation of relevant OSHA standards, see the OSHA hexavalent chromium topic page (http://www.osha.gov/SLTC/hexavalent-chromium/index.html).

8.1 NIOSH Recommended Exposure Limit

8.1.1 The NIOSH REL

NIOSH recommends that airborne exposure to all Cr(VI) compounds be limited to a concentration of 0.2 µg Cr(VI)/m^3 for an 8-hr TWA exposure during a 40-hr workweek. The use of NIOSH Method 7605 (or validated equivalents) is recommended for Cr(VI) determination. The REL represents the upper limit of exposure for each worker during each work shift. Because of the residual risk of lung cancer at the REL, NIOSH further recommends that all reasonable efforts be made to reduce exposures to Cr(VI) compounds below the REL. The available scientific evidence supports the inclusion of all Cr(VI) compounds into this recommendation. The REL is intended to reduce workers' risk of lung cancer associated with occupational exposure to Cr(VI) compounds over a 45-year working lifetime. It is expected that reducing airborne workplace exposures will also reduce the nonmalignant respiratory effects of Cr(VI) compounds, including irritated, ulcerated, or perforated nasal septa and other potential adverse health effects.

In addition to limiting airborne concentrations of Cr(VI) compounds, NIOSH recommends that dermal exposure to Cr(VI) be prevented in the workplace to reduce the risk of adverse dermal health effects, including irritation, ulcers, allergic contact dermatitis, and skin sensitization.

8.1.2 Sampling and Analytical Methods

The sampling and analysis of Cr(VI) in workplace air should be performed using precise, accurate, sensitive, and validated methods. The use of NIOSH Method 7605 (or validated equivalents) is recommended for Cr(VI) determination in the laboratory. Other standardized methods for Cr(VI) analysis include OSHA Method ID-215 [OSHA 1998, 2006], ASTM Method D6832-02 [ASTM 2002], and ISO Method 16740 [ISO 2005]. More detailed discussion of sampling and analytical methods for Cr(VI) is provided in Chapter 3, "Measurement of Exposure."

8.2 Informing Workers about the Hazard

8.2.1 Safety and Health Programs

Employers should establish a comprehensive safety and health training program for all workers who manufacture, use, handle, or dispose of Cr(VI) compounds, or perform any other activity that involves exposure to Cr(VI) compounds. This program should include training on workplace hazards, monitoring of airborne Cr(VI) levels, and medical surveillance of employees exposed to Cr(VI).

Workers should receive training as mandated by the revised OSHA Hazard Communication Standard (HCS, 29 CFR 1910.1200). OSHA revised the HCS to align with the United Nations' Globally Harmonized System of Classification and Labeling of Chemicals (GHS). Employers should be aware of the changes, requirements, phase-in dates, and compliance effective dates of the revised HCS. Workers should be trained on the new label elements (pictograms, signal words, hazard statements, and precautionary statements) and Safety Data Sheet (SDS) format.

Worker training should include information about the following: the Cr(VI) compounds to which they are exposed; the physical and chemical properties of these compounds; explanation of material safety data sheets (MSDSs), SDSs, and label elements; appropriate routine and emergency handling procedures; and recognition of the adverse health effects of Cr(VI) exposure. Training should be provided about the industrial hygiene hierarchy of controls, how to implement controls to prevent and reduce exposures, and the appropriate use, maintenance, and storage of PPE to minimize Cr(VI) exposure. Workers should be trained to report promptly to their supervisor any leaks observed, failures of equipment or procedures, wet or dry spills, cases of gross contact, and instances of suspected overexposure to Cr(VI) compounds.

Employees should be trained to report to their supervisor or the director of the medical monitoring program any symptoms or illnesses associated with Cr(VI) exposure and any workplace events involving accidental or incidental exposures to Cr(VI) compounds. A medical monitoring program should be in place for workers exposed to Cr(VI) compounds in the workplace (see Section 8.6).

Safety and health programs should also include workers involved in cleaning, repair, and maintenance procedures that may involve exposure to Cr(VI) compounds. When possible, these duties should be performed when the work area or facility is not in operation to minimize these workers' airborne and dermal Cr(VI) exposures.

8.2.2 Labeling and Posting

Receptacles containing Cr(VI) compounds used or stored in the workplace should carry a permanently attached label that is readily visible. The label should identify Cr(VI) compounds and provide information about their adverse health effects, including cancer, and appropriate emergency procedures. Labels should meet the requirements of the revised OSHA HCS (29 CFR 1910.1200).

Signs containing information about the health effects of Cr(VI) compounds should be posted at the entrances to work areas or building enclosures and in visible locations throughout the work areas where there is a potential for exposure to Cr(VI) compounds. Because Cr(VI) compounds are carcinogenic, the following warning sign, or a sign containing comparable information that is consistent with the workplace hazard communication program, should be posted:

> DANGER; CHROMIUM(VI); MAY CAUSE CANCER; CAN DAMAGE SKIN, EYES, NASAL PASSAGES, AND LUNGS; AUTHORIZED PERSONNEL ONLY

In areas where respirators and/or chemical protective clothing are needed, the following statement should be added:

> RESPIRATORY PROTECTION AND CHEMICAL PROTECTIVE CLOTHING REQUIRED IN THIS AREA

Information about emergency first-aid procedures and the locations of emergency showers and eyewash fountains should be provided where needed.

All signs should be printed both in English and in the language(s) of non-English-speaking workers. All workers who are unable to read should receive oral instruction on the content and instructions on any written signs. Signs using universal safety symbols should be used wherever possible.

8.3 Exposure Control Measures

Many exposure control measures are used to protect workers from potentially harmful exposures to hazardous workplace chemical, physical, or biological agents. These control measures include, in order of priority: elimination, substitution, engineering controls, administrative controls and appropriate work practices, and the use of protective clothing and equipment [NIOSH 1983b]. The occupational exposure routes of primary concern for Cr(VI) compounds are the inhalation of airborne Cr(VI) and direct skin contact. This section provides information on general exposure control measures that can be used in many workplaces and specific control measures for controlling Cr(VI) exposures that are effective in some workplaces.

8.3.1 Elimination and Substitution

Elimination of a hazard from the workplace is the most effective control to protect worker health. Elimination may be difficult to implement in an existing process; it may be easier to implement during the design or re-design of a product or process.

If elimination is not possible, substitution is the next choice of control to protect worker health. Using substitution as a control measure may include substitution of equipment, materials, or less hazardous processes. Equipment substitution is the most common type of substitution [AIHA 2011; NIOSH 1973b]. It is often less costly than process substitution, and it may be easier than finding a suitable substitute material. An example that applies to Cr(VI) exposure reduction is the substitution of an enclosed and automated spray paint booth for a partially enclosed workstation.

Material substitution is the second most common type of substitution after equipment substitution. It has been used to improve the safety of a process or lower the intrinsic toxicity of the material being used. However, evaluation of the potential adverse health effects of the substitute material is essential to ensure that one hazard is not replaced with a different one [AIHA 2011; NIOSH 1973b].

Blade et al. [2007] reported material substitution in some processes with potential worker exposures to Cr(VI) compounds investigated by NIOSH between 1999 and 2001. A reduction in the use of chromate-containing paints was reported in construction (i.e., bridge repainting) and vehicle manufacturing (i.e., the manufacture of automobiles and most trucks reportedly no longer uses chromate paints). However, chromate-containing paints reportedly remained without satisfactory substitute in aircraft manufacture and refurbishing. Chromium electroplating industry representatives also reported steady demand for hard chrome finishes for mechanical parts such as gears, molds, etc., because of a lack of economical alternatives for this durable finish.

Many examples of process substitution have been considered. A change from an intermittent or batch-type process to a continuous-type process often reduces the potential hazard, particularly if the latter process is more automated [AIHA 2011; NIOSH 1973b; Soule 1978]. Dipping objects into a coating material, such as paint, usually causes less airborne material and is less of an inhalation hazard than spraying the material.

Reducing the Cr(VI) Content of Portland Cement. One example of substitution is using Portland cement with a reduced Cr(VI) content to reduce workers' risk of skin sensitization. The trace amount of Cr(VI) in cement can cause allergic contact dermatitis that can be debilitating and marked by significant, long-term adverse effects [NIOSH 2005a]. The chromium in cement can originate from a variety of sources, including raw materials, fuel, refractory brick, grinding media, and additions [Hills and Johansen 2007]. The manufacturing process, including the kiln conditions, determines how much Cr(VI) forms. The Cr(VI) content of cement can be lowered by using materials with lower chromium content during production and/or by adding agents that reduce Cr(VI). The use of slag, in place of or blended with clinker, may decrease the Cr(VI) content [Goh and Gan 1996; OSHA 2008]. Ferrous sulfate is the material most often added to cement to reduce its Cr(VI) content.

Since 2005, the European Union has restricted cement and cement-containing products with potential skin contact to a limit of 2 ppm soluble Cr(VI) [European Parliament and the Council of the European Union 2003]. Reducing the Cr(VI) content of cement has resulted in a reported decrease in the number of cases of allergic contact dermatitis [Avnstorp 1989; Geier et al. 2010; Roto et al. 1996]. Limiting the Cr(VI) content of cement in the United States warrants consideration. Further research on the potential impacts of this change in U.S. industry is needed.

8.3.2 Engineering Controls

If elimination or substitution are not possible, engineering controls are the next choice for reducing worker exposure to Cr(VI) compounds. These controls should be considered when new facilities are being designed or when existing facilities are being renovated to maximize their effectiveness, efficiency, and economy. Engineering measures to control potentially hazardous workplace exposures to Cr(VI) compounds include isolation and ventilation. OSHA determined that the primary engineering control measures most likely to be effective in reducing employee exposure to airborne Cr(VI) are local exhaust ventilation (LEV), process enclosure, process modification, and improved general dilution ventilation [71 Fed. Reg. 10099 (2006)]. These and other engineering controls are described in the following sections.

8.3.2.1 Isolation

Isolation as an engineering control may involve the erection of a physical barrier between the worker and the hazard. Isolation may also be achieved by the appropriate use of distance or time [Soule 1978]. Examples of hazard isolation include the isolation of potentially hazardous materials into separate structures, rooms, or cabinets; and the isolation of potentially hazardous process equipment into dedicated areas or rooms that are separate from other work areas [AIHA 2011; NIOSH 1973b]. Separate ventilation of the isolated area(s) may be needed to maintain the isolation of the hazard from the rest of the facility [Soule 1978]. Complete isolation of an entire process also may be achieved using automated, remote operation methods [AIHA 2011; NIOSH 1973b].

An example of an isolation technique to control Cr(VI) exposure is the use of a separate, ventilated mixing room for mixing batches of powdered materials containing chromate pigments.

8.3.2.2 Ventilation

Ventilation may be defined as the strategic use of airflow to control the environment within a space—to provide thermal control within the space, remove an air contaminant near its source of release into the space, or dilute the concentration of an air contaminant to an acceptable level [Soule 1978]. When controlling a workplace air contaminant such as Cr(VI), a specific dedicated exhaust ventilation system or assembly might need to be designed for the task or process [AIHA 2011; NIOSH 1973b].

Local exhaust ventilation (LEV) is primarily intended to capture the contaminant at specific points of release into the workroom air through using exhaust hoods, enclosures, or similar assemblies. LEV is appropriate for the control of stationary point sources of contaminant release. It is important to assure proper selection, maintenance, placement, and operation of LEV systems to ensure their effectiveness [ACGIH 2010].

General ventilation, often called dilution ventilation, is primarily intended to dilute the concentration of the contaminant within the general workroom air. It controls widespread problems such as generalized or mobile emission sources [AIHA 2011; NIOSH 1973b]. Whenever practicable, point-source emissions are most effectively controlled by LEV, which is designed to remove the contaminant at the source before it emanates throughout the workspace. Dilution ventilation is less effective because it merely reduces the concentration of the contaminant after it enters the workroom air, rather than preventing much of the emitted contaminant from ever entering the workroom air. It also is much less efficient, requiring much greater volumetric airflow to reduce concentrations. However, for non-point sources of contaminant emission, dilution ventilation may be necessary to reduce exposures.

The air exhausted by a LEV system must be replaced, and the replacement air will usually be supplied by a make-up air system that is not associated with any particular exhaust inlet and/or by simple infiltration through building openings (relying on infiltration for make-up air is not recommended). This supply of replacement air will provide general ventilation to the space even if all the exhaust is considered local. The designation of a particular ventilation system or assembly as local or general, exhaust or supply, is governed by the primary intent of the design [AIHA 2011; NIOSH 1973b].

Push-pull ventilation may be used to control exposures from open surface tanks such as electroplating tanks. Push-pull ventilation includes a push jet located on one side of a tank and a lateral exhaust hood on the other side [ACGIH 2010]. The jet formed over the tank surface captures the emissions and carries them into the hood. Many other types of ventilation systems may be used to control exposures in specific workplace operations [ACGIH 2010].

8.3.2.3 Examples of engineering controls to reduce Cr(VI) exposures

Many types of engineering controls have been used to reduce workplace Cr(VI) exposures. Some of the engineering controls recommended by NIOSH in 1975 [NIOSH 1975a] are still valid and in use today. Some examples are included here, there are many others [ACGIH 2010; AIHA 2011].

Closed systems and operations can be used for many processes, but it should be ensured that seals, joints, covers, and similar assemblies fit properly to maintain negative static pressure within the closed equipment, relative to the surroundings.

The use of LEV may be needed even with closed systems to prevent workers' exposures during operations such as unloading, charging, and packaging. The use of protective clothing and equipment may also be needed. Ventilation systems should be regularly inspected and maintained to assure effective operation. Work practices that may obstruct or interfere with ventilation effectiveness must be avoided. Any modifications or additions to the ventilation system should be evaluated by a qualified professional to ensure that the system is operating at design specifications.

The use of clean areas, such as control rooms supplied with uncontaminated air, is one method of isolating the workers from the hazard. An area to which workers may retreat for periods of time when they are not needed at the process equipment also may be configured as a clean area.

The most difficult exposures to control often are those of repair and maintenance workers who may be working in emergency conditions in close contact with contaminated equipment or surfaces. Their exposures may be variable in nature and irregular in frequency. Controls such as LEV should be used where practicable, but the use of PPE may be required when the use of engineering controls is not feasible or as effective as necessary.

From 1999 through 2001, NIOSH conducted field surveys in 21 workplaces across a variety of industrial operations and industry sectors with potential worker exposures to Cr(VI) compounds [Blade et al. 2007]. Many of the observed processes and equipment applications were typical of those throughout industry, such as dip tanks, paint booths, and grinding, sanding, and welding operations. The application of general or specialized engineering controls were observed or recommended to control exposures in these operations.

The following are examples of industry processes or job tasks where different types of engineering control measures can be applied to reduce Cr(VI) exposures. The exposure data from these NIOSH field surveys is presented in Tables 2–4 through 2–7 (see Chapter 2).

Chromium electroplating. A combination of engineering measures may be needed to effectively control potential exposures during chromium electroplating processes, including hard chrome plating. Hard chrome is a relatively thick coating of chromium that provides an extremely durable, wear-resistant surface for mechanical parts. At one facility studied by NIOSH, push-pull ventilation systems, polyethylene tarpaulins, and a foam-blanket mist-suppressant product were used, but workers' exposures still exceeded the existing NIOSH REL of 1 µg Cr(VI)/m^3 [Blade et al. 2007]. Qualitative airflow visualization using smoke tubes suggested that the push-pull ventilation systems were generally effective in moving air away from workers' breathing zones. However, maintenance problems with the ventilation system suggested that the system was not always operating effectively. Floating plastic balls had reportedly been used in the past but proved impractical. Mist suppressants that reduce surface tension were not used because of concerns that they may induce pitting in the hard chrome-plated finish.

In contrast with hard chrome plating tanks, control of bright chrome plating tank emissions is less problematic. Bright chrome plating provides a thin chromium coating for appearance and corrosion protection to non-mechanical parts. The use of a wetting agent as a fume suppressant that reduces surface tension provided effective control of emissions [Blade et al. 2007].

At another facility, a hard chrome-plating tank was equipped with a layer of a newly developed, proprietary viscous liquid and a system to circulate it [Blade et al. 2007]. This system effectively reduced mist emission containing Cr(VI) from the tank, but it was not durable.

Welding and thermal cutting involving chromium-containing metals. Many welding task variables affect the Cr(VI) content of welding fume and the associated Cr(VI) exposures. Both the base metal of the parts being joined and the consumable metal (welding rod or wire) added to create the joint have varying compositions of chromium. The welding process and shield-gas type, and the Cr content of both the consumable material and the base metal affect the Cr(VI) content of the fume [Keane et al. 2009; Heung et al. 2007; EPRI 2009; Meeker et al. 2010]. When possible, process and material substitution may be effective in reducing welding Cr(VI) exposures. Evaluation of an exposure database from The Welding Institute indicated that welding of stainless steel or Inconel (a nickel-chromium alloy containing 14–23% Cr[VI]) resulted in median Cr(VI) exposures of 0.6 µg/m^3 compared with the Cr(VI) exposures of the welding of other metals, which were less than the LOD (range 0.1–0.2 µg/m^3) [Meeker et al. 2010]. Processes such as gas-tungsten arc welding (GTAW), submerged arc welding (SAW), and gas-metal arc welding (GMAW) tend to generate less fume [Fiore 2006]. Whenever appropriate, the selection of GTAW will help to minimize Cr(VI) exposures in welding fume [EPRI 2009]. Cr(VI) exposures during shielded-metal arc welding (SMAW) may be minimized by using consumables (welding rod or wire) containing low chromium content (i.e., less than 3% Cr) [EPRI 2009].

Welding or thermal-cutting fumes containing Cr(VI) are often controlled using LEV systems [Blade et al. 2007]. Two common LEV systems are high-volume low-vacuum systems or low-volume high vacuum systems. High-volume low-vacuum systems have large-diameter hoses or ducts that result in larger capture distances [Fiore 2006]. High-vacuum low-volume systems use smaller hoses and so have a smaller capture distance; they are often more portable [Fiore 2006]. In controlled welding trials, the use of a portable high-vacuum fume extraction unit reduced Cr(VI) exposures from a median of 1.93 µg/m^3 to 0.62 µg/m^3 ($P = 0.02$) [Meeker et al. 2010].

When welding outdoors, the effect of the wind and the position of the welder are important factors controlling the effectiveness of LEV [NIOSH 1997]. In the field setting LEV effectiveness is directly related to proper usage [Meeker et al. 2010]. Proper positioning of the ventilation inlet relative to the welding nozzle and the worker's breathing zone is critical to exposure-control performance; this often requires frequent repositioning by the welder [Fiore 2006]. Welders may keep the LEV inlet too far from the weld site to be effective, or they may be reluctant to use the LEV system because of concerns that the incoming ventilation air could adversely affect the weld quality by impairing flux or shield-gas effectiveness [EPRI 2009; Fiore 2006; Meeker et al. 2010].

Specialized systems called "fume extraction welding guns" can be used in many workplaces (e.g., outdoors) to reduce worker exposure to welding fumes. These systems combine the arc-welding gun with a series of small LEV air inlets so that the air inlets are always at a close distance to the welding arc. These systems are heavier and more cumbersome than standard arc-welding guns, so ergonomic issues must be considered [Fiore 2006].

Spray application of chromate-containing paints. Blade et al. [2007] determined that the most effective measure for reducing workers' Cr(VI) exposures at a facility where chromate-containing paints were applied to aircraft parts would be the substitution of paints with lower chromate content (i.e., 1% to 5%) for those with higher content (i.e., 30%) wherever possible [Blade et al. 2007]. Results indicated that partially enclosed paint booths for large-part painting might not provide adequate contaminant capture. The facility also used fully enclosed paint booths with single-pass ventilation, with air entering one end and exhausted from the other. The average internal air velocities within these booths needed to exceed the speed with which the workers walked while spraying paint so that the plume of paint overspray moved away from the workers.

Removal of chromate-containing paints. At a construction site where a bridge was to be repainted, the removal of the existing chromate-containing paint was accomplished by abrasive blasting. An enclosure of plastic sheeting was constructed to contain the spent abrasive and paint residue and prevent its release into the surrounding environment [Blade et al. 2007]. No mechanical ventilation was provided to the containment structure. NIOSH recommended that this type of containment structure be equipped with general-dilution exhaust ventilation that discharges the exhausted air through a high-efficiency particulate air (HEPA) filtration unit.

Other types of specialized engineering measures applicable for the control of exposures during chromate-paint removal have been investigated and recommended for selected applications. These recommendations are often made in the context of lead exposure control [OSHA 1999b], but they are relevant to Cr(VI) control because lead chromate paints may be encountered during paint removal projects. Such control measures include high-pressure water blasting, wet-abrasive blasting, vacuum blasting, and the use of remotely controlled automated blasting devices [Meeker et al. 2010]. High-pressure water blasting uses a blast of extremely focused water at high velocity to remove paint and corrosion, but it does not reprofile the underlying metal substrate for repainting. Wet-abrasive blasting uses a conventional blasting medium that is wetted with water to remove the paint and corrosion and to reprofile the metal. The wetted medium helps suppress the emission of dust that contains removed chromate-paint particles. Vacuum blasting uses a blasting nozzle surrounded by a vacuum shroud with a brush-like interfacing surface around its opening, which the operator keeps in contact with the metal surface being blasted. Large reductions in exposures have been reported with this system, but considerations include the following: good work practices are needed to assure proper contact with the surface is maintained; the full assembly is heavier than conventional nozzles and thus raises ergonomic concerns; and production (removal) rates reportedly are much lower than with conventional blast nozzles [Meeker et al. 2010].

Mixing of chromate-containing pigments. At a colored-glass manufacturing facility, pigments containing Cr(VI) were weighed in a separate room with LEV, then moved to a production area for mixing into batches of materials [Blade et al. 2007]. Cr(VI) exposures at the facility were very low to not detectable.

At a screen printing ink manufacturing facility, there was no dedicated pigment-mixing room; LEV was used at the ink-batch mixing and weighing operation, but capture velocities were inadequate [Blade et al. 2007]. Almost all the Cr(VI) exposures of the ink-batch weighers exceeded the existing REL.

Operations creating concrete dust. Portland cement contains Cr(VI), so operations that create concrete dust have the potential to expose workers to Cr(VI). In one operation studied by NIOSH, the use of water to suppress dust during cleanup was observed to result in visibly lower dust concentrations [Blade et al. 2007]. All Cr(VI) exposures at the facility were low. At a construction-rubble crushing and recycling facility, a water-spray system was used on the crusher at various locations, and the operator also used a hand-held water hose [Blade et al. 2007]. All Cr(VI) exposures at this facility also were low.

The exposure data from these examples from the NIOSH field surveys are presented in Tables 2–4 through 2–7 (see Chapter 2).

8.3.3 Administrative Controls and Work Practices

Administrative controls are changes in work practices and management policies designed to minimize exposure times. Appropriate work practices include proper techniques to handle materials, good personal hygiene and sanitation practices, and good housekeeping in the work area. Employers should ensure that water and soap are available to promote good hand hygiene practices.

Work areas with potential Cr(VI) exposures should have restricted access so that only those workers assigned to and trained for the task or process are allowed to enter. Workers should not be allowed to smoke, eat, or drink in work areas where Cr(VI) compounds are used or stored. Emergency showers and eye-flushing fountains should be provided by the employer in areas with the potential for skin or eye contact with Cr(VI). This equipment should be properly maintained and inspected regularly. If Cr(VI) gets on the skin, the affected area must be flushed promptly with large amounts of mild soap and running water for at least 15 minutes. If the eyes are contaminated with Cr(VI), they should be flushed immediately for at least 15 minutes with a copious flow of water and promptly examined by a physician.

Clean work clothing should be put on before each work shift. The clothing should be changed whenever it becomes wetted or grossly contaminated with compounds containing Cr(VI). Work clothing should not be worn home. Workers should be provided with showering and changing areas free from contamination, where they may store and change into street clothes before leaving the worksite. Employers should provide services for laundering work clothing so that contaminated clothes are not taken home. These precautions will protect the worker and people outside the workplace, including the worker's family, from being exposed to clothing contaminated with Cr(VI). Laundry personnel should be informed about the potential hazards of handling contaminated clothing and should be instructed about measures to minimize their health risk.

8.3.3.1 Portland cement work practices

Clean water, non-alkaline soap, and clean towels should be available for workers exposed to Portland cement [OSHA 2008]. Wet clean-up methods (e.g., hose, then squeegee or mop) should be used rather than dry sweeping. The surfaces of all tools should be cleaned after use. Additional recommended work practices for preventing dermal exposure to Portland cement are provided by OSHA [2008].

8.3.4 Protective Clothing and Equipment

The use of protective clothing and PPE is another way to create a physical barrier between the worker and the hazard. It may be appropriate to use different types of protective clothing and PPE, such as chemically impervious gloves, clothing, and respirators. Employers are responsible for the selection of PPE, training in the proper use of PPE, ensuring the PPE is properly used, maintenance of PPE, and providing and paying for all PPE [NIOSH 1999]. The use of respirators to control inhalation exposures to air contaminants is considered a last resort for cases where engineering and other measures cannot provide sufficient control. Workers should be trained in the proper use, maintenance, and storage of all protective clothing worn in the workplace.

Workers and persons responsible for worker health and safety should be informed that protective clothing might interfere with the body's heat dissipation, especially during hot weather or in hot work situations. Additional monitoring is required to prevent heat-related illness when protective clothing is worn under these conditions [NIOSH 1986].

8.3.4.1 Protective clothing and gloves

NIOSH recommends the use of gloves, eye protection, and chemical protective clothing (CPC) for workers with potential skin or eye contact with Cr(VI) compounds. Dermal and mucous membrane contact with Cr(VI) compounds should be prevented by full-body protective clothing consisting of the following: head, neck, and face protection; coveralls or similar protective body clothing; impermeable gloves with gauntlets; and shoes and apron where solutions or dry materials containing Cr(VI) may be contacted.

Protective clothing and gloves made from PVC or Saranex can be used for an 8-hour exposure, while those made from butyl or Viton can be used for a 4-hour exposure [Forsberg and Keith 1999]. Although the selection of this CPC is based on permeation properties, other selection factors, including size, dexterity, cut and tear resistance, and glove use with other chemicals should be considered. Contaminated CPC, gloves, and shoes should be discarded or decontaminated with proper methods before reuse. If Cr(VI) gets on the skin, the affected area must be flushed immediately with large amounts of mild soap and running water for at least 15 minutes.

The proper use of protective clothing requires that all openings be closed and that all garments fit snugly about the neck, wrists, and ankles whenever the wearer is in an exposure area. Care must be exercised to keep work clothing separate from street clothing to avoid contamination. All protective clothing must be maintained properly in an uncontaminated environment. Protective clothing should be inspected before each use and cleaned or replaced regularly.

Eye protection should be provided by the employer and used by the employees where eye contact with Cr(VI) is possible. Selection, use, and maintenance of eye protective equipment should be in accordance with the provisions of the American National Standard Practice for Occupational and Educational Eye and Face Protection, ANSI Z87.1-1989 [ANSI 1989]. In work environments where Cr(VI) levels are above the NIOSH REL and respiratory protection is required, NIOSH recommends that eye protection be incorporated into PPE by the use of tight-fitting full facepiece respirators, or tight-fitting half-mask respirators used in conjunction with safety spectacles or goggles.

See Section 8.3.4.3 for PPE recommendations for workers with dermal contact with Portland cement. Further information on chemical protective clothing is available on the NIOSH Protective Clothing and Ensembles topic page (http://www.cdc.gov/niosh/topics/protclothing) and in the OSHA Technical Manual, Section VIII, Chapter 1, "Chemical

Protective Clothing" [OSHA 1999a] (http://www.osha.gov/dts/osta/otm/otm_viii/otm_viii_1.html).

8.3.4.2 Respiratory protection

NIOSH recommends respirator use while performing any task for which the airborne exposure concentration is unknown or has been documented to be higher than the NIOSH REL of 0.2 µg Cr(VI)/m^3 8-hr TWA. Respirators should not be used as the primary means of controlling worker exposures. Other exposure control methods such as elimination, substitution, engineering controls, administrative controls, and changes in work practices should be implemented in an attempt to keep exposures before the REL before the use of respirators is required. However, the use of respirators may be necessary when the airborne exposure concentration is unknown or when other control measures do not control airborne Cr(VI) concentrations to below the REL. NIOSH recognizes this may be a particular challenge in electroplating, spray painting, atomized-alloy spray-coating operations, some types of welding operations, and other industries or tasks with routinely or uncontrollably high Cr(VI) exposures.

When respiratory protection is needed, the employer should establish a comprehensive respiratory protection program as described in the OSHA respiratory protection standard [29 CFR 1910.134]. Elements of a respiratory protection program, established and described in a written plan that is specific to the workplace, must include the following:

- Procedures for selecting respirators.
- Medical evaluations of employees required to wear respirators.
- Fit-testing procedures.
- Routine-use procedures and emergency respirator use procedures.
- Procedures and schedules for cleaning, disinfecting, storing, inspecting, repairing, discarding, and maintaining respirators.
- When applicable, procedures for ensuring adequate air quality for supplied air respirators (respirable air should meet the requirements of Compressed Gas Association Specification G-7.1 Grade D or higher quality).
- Training in respiratory hazards.
- Training in proper use and maintenance of respirators.
- Program evaluation procedures.
- Procedures for ensuring that workers who voluntarily wear respirators (excluding filtering-facepiece respirators) comply with the medical evaluation and cleaning, storing, and maintenance requirements of the standard.
- A designated program administrator who is qualified to administer the respiratory protection program.

The written program should be updated as necessary to account for changes in the workplace that affect respirator use. All equipment, training, and medical evaluations required under the respiratory protection program should be provided at no cost to workers.

NIOSH recommends that the selection of respiratory protection follow the guidance in NIOSH Respirator Selection Logic [NIOSH 2004]. Table 8–1 provides respirator selection recommendations for Cr(VI). A comprehensive assessment of all workplace exposures should be performed to determine the presence of other possible contaminants to ensure that the proper respiratory protection is used.

Further information about respiratory protection is available on the OSHA Respiratory Protection topic page (http://www.osha.gov/SLTC/respiratoryprotection/index.html), the

NIOSH Respirators topic page (http://www.cdc.gov/niosh/topics/respirators/), the *NIOSH Guide to Industrial Respiratory Protection* [NIOSH 1987a], the *NIOSH Guide to the Selection and Use of Particulate Respirators Certified Under 42 CFR 84* [NIOSH 1996b], and NIOSH Respirator Selection Logic [2004].

8.3.4.3 PPE for Portland cement exposure

Employers should provide PPE, including gloves, boots, and eye protection for workers exposed to Portland cement [OSHA 2008]. Butyl or nitrile gloves are often recommended for caustic materials such as Portland cement. Cotton or leather gloves are not recommended. Employers should consult the cement manufacturer's MSDS for information about the proper gloves to provide. Sturdy, slip-resistant, waterproof boots should be provided when needed to prevent wet cement from contacting workers' skin. OSHA [2008] provides additional guidance on preventing skin problems from working with Portland cement.

8.4 Emergency Procedures

Emergency plans and procedures should be developed for all work areas where there is a potential for exposure to Cr(VI). Workers should be trained in the effective implementation of these plans and procedures. These plans should be reviewed regularly for their effectiveness and updated when warranted because of changes in the facility, operating procedures, or chemical types or uses. Necessary emergency equipment, including appropriate respiratory protective devices, should be kept in readily accessible locations. Appropriate respirators should be available for use during evacuation. A full facepiece respirator with a 100-level filter or any appropriate escape-type, self-contained breathing apparatus should be used for escape-only situations. If chromyl chloride is present, a full facepiece gas mask (14G) with canister providing organic vapor (OV) and acid gas (AG) protection with a 100-level filter or any appropriate escape-type, self-contained breathing apparatus should be used for escape-only situations.

Any spills of Cr(VI) compounds should be promptly cleaned up by means that minimize the inhalation of, or contact with, the spilled material. No dry sweeping should be performed. Wet vacuuming is preferred for spills of dry material. Wet spills and flushing of wet or dry spills should be channeled for appropriate treatment or collection for disposal. They should not be channeled directly into the sanitary sewer system. Dry vacuuming is acceptable only if an adequately filtered system is used—either a HEPA-filtered system or a single-pass externally exhausted system.

8.5 Exposure Monitoring Program

The employer should establish a workplace program to monitor exposure to airborne Cr(VI). The program should include environmental and personal monitoring of airborne Cr(VI) exposure concentrations and focus on identifying potential exposures and assessing the effectiveness of exposure controls. The goal of the exposure monitoring program is to ensure a more healthful work environment where worker exposures do not exceed the REL. The exposure monitoring program should systematically (1) characterize the exposures of all workers; (2) identify those workers, processes, and tasks with the highest exposures; (3) identify processes or tasks where worker exposures exceed the REL; (4) assess the effectiveness of engineering controls and work practices; and (5) determine the need for PPE use.

Historically, NIOSH has recommended an action level (AL) with the primary consideration of protecting workers from exposures above the REL [NIOSH 1975b]. Exposure concentrations measured at or above the AL were thought to indicate with a high degree of certainty that exposure concentrations exceeded the REL, which triggered additional controls and administrative actions to reduce worker exposures. Although the term "action level" was not used in the 1975 NIOSH Cr(VI) criteria document, occupational exposure to carcinogenic Cr(VI) was defined as "exposure to airborne Cr(VI) at concentrations greater than one-half of the workplace environmental limit for carcinogenic Cr(VI)" [NIOSH 1975a].

NIOSH is re-evaluating its policy of recommending an AL set at one-half the REL. Cr(VI) exposures are highly variable within and across diverse workplaces. Because of the great range and high variability of Cr(VI) exposures across workplaces, it is not possible to establish a specific AL for Cr(VI) compounds. Therefore, NIOSH is providing general exposure monitoring guidance for workplaces with Cr(VI) exposures rather than recommending one specific AL for all Cr(VI) compounds. This will allow each employer to determine a strategy to monitor exposure specific to each workplace that assures that worker exposures do not exceed the REL.

A strategy to monitor exposure should be developed and implemented for each specific process and group of workers exposed to Cr(VI) compounds. The details of the plan will depend on a number of factors, including the number of workers in the group and variability in exposure within the group. Workers' airborne exposures vary from day to day, and the daily exposures are typically log normally distributed. Exposures in well-controlled processes and environmental conditions vary less than in poorly controlled processes and where the environmental conditions change considerably, such as outdoors. Greater day-to-day variability of 8-hr TWA exposures requires that more daily 8-hr exposures be assessed to achieve the specified level of confidence in the sampling results.

The strategy to monitor exposure should provide enough information to adequately describe the distribution of workers' exposures. It should include exposure-sampling surveys that produce a high degree of confidence that workers' daily 8-hr TWA exposures are maintained below the REL. As part of the initial workplace hazard surveillance, the workers, tasks, and processes associated with the highest Cr(VI) exposures should be identified. A sampling strategy that focuses on these workers with the highest perceived exposure concentrations may be more practical than a random sampling approach. A focused sampling strategy is most efficient for identifying exposures above the REL if maximum-risk workers and time periods are accurately identified [NIOSH 1977; Leidel and Busch 1994]. The exposure-sampling survey should be performed by collecting representative personal samples over the entire work shift. Whenever possible, personal samples should be collected in the breathing zone of the worker. Periodic exposure monitoring should be performed at least annually, and whenever any major process change takes place or there is another reason to suspect that exposure concentrations may have changed. All workers should be notified of the results of their exposure monitoring and of any actions taken to reduce their exposure. More detailed information on developing exposure-monitoring plans for specific situations is available from NIOSH [1977] and the AIHA [2006].

For workers exposed to concentrations of Cr(VI) compounds above the REL, an evaluation of existing control measures should be conducted to determine if the measures are operating as expected. If existing control measures are inadequate, then additional control measures should be implemented to reduce 8-hr TWA exposures to below the REL. After controls are implemented, an exposure-sampling survey should be performed that can produce a high degree of confidence that daily 8-hr TWA exposures are below the REL [NIOSH 1977; Leidel and Busch 1994]. Periodic exposure monitoring of workers should continue to ensure that workers' exposures are maintained below the REL.

NIOSH Method No. 7605 (or validated equivalents) should be used for the collection and analysis of airborne Cr(VI) samples in the workplace. Area sampling (environmental) may be useful to determine sources of airborne Cr(VI) exposures and assessing the effectiveness of engineering controls. Important air sampling considerations, including the possible reduction of Cr(VI) to Cr(III) during sampling and sample preparation, are presented in Section 3.1.2.

The exposure monitoring program should also include the assessment and monitoring of potential dermal exposure associated with any area, task, or process. Where there is the potential for dermal exposure to Cr(VI) compounds, controls should be implemented or administrative actions should be taken (e.g., PPE) to reduce potential exposures.

8.6 Medical Monitoring

The employer should establish a medical monitoring program for workers with occupational exposure to Cr(VI) compounds, including personnel involved with routine or emergency repair or maintenance, as specified in Section 8.6.1 below. Medical monitoring represents secondary prevention and should not replace the primary prevention efforts mentioned in previous sections of this chapter to minimize occupational exposure to Cr(VI). The goal of a workplace medical monitoring program is the early identification of adverse health effects that may be related to Cr(VI) exposure, such as dermatitis, respiratory irritation, airway obstruction and other local or systemic effects. It is hoped that early detection of adverse health effects, subsequent treatment, and workplace interventions will minimize the adverse health effects of Cr(VI) exposure. Medical monitoring data may also be used for the purposes of medical surveillance to identify work areas, tasks, and processes that require additional primary prevention efforts.

8.6.1 Worker Participation

Workers potentially exposed to Cr(VI) compounds as specified below may benefit by being included in an occupational medical monitoring program. Workers should be provided with information about the purposes of medical monitoring, the health benefits of the program, and the procedures involved. When possible, employers or the designated program director should characterize worker exposures to identify potential airborne and/or dermatologic Cr(VI) exposures. The following hierarchy describes workers with airborne Cr(VI) exposure who should be included in a medical monitoring program and could receive the greatest benefit from medical screening:

- Workers, including those using respiratory protective equipment, in workplace environments where airborne Cr(VI) concentrations cannot be controlled below the REL of 0.2 $\mu g/m^3$.
- Workers with unintended but potentially high airborne Cr(VI) exposures during situations such as emergencies or PPE failure.
- Workers exposed to Cr(VI), regardless of airborne Cr(VI) concentration, who develop signs, symptoms, or respiratory changes apparently related to Cr(VI) exposure.
- Workers exposed to Cr(VI) in their current job who may have been previously exposed to asbestos or other respiratory hazards that place them at an increased risk of respiratory disease.
- Workers with dermal exposure to Cr(VI) compounds.

Workers exposed to Cr(VI) by the dermatologic route (either with or without accompanying airborne Cr[VI] exposure) are at risk of adverse health effects such as dermatitis and other local effects and should be included in a medical monitoring program even if not exposed to Cr(VI) airborne concentrations above the REL. These workers may not need to be included in the respiratory effects portion of the monitoring program if they have no potential risk of airborne Cr(VI) exposure.

8.6.2 Medical Monitoring Program Director

The employer should assign responsibility for the medical monitoring program to a qualified physician or other qualified health care provider (as determined by appropriate state laws and regulations) who is informed and knowledgeable about the following:

- Administration and management of a medical monitoring program for occupational hazards.
- Establishment of a respiratory protection program, based on an understanding of the requirements of the OSHA respiratory protection standard and types of respiratory protection devices available at the workplace.
- Identification and management of occupational skin disease.
- Identification and management of occupational respiratory effects or illnesses, including lung cancer.

8.6.3 Medical Monitoring Program Elements

Recommended elements of a medical monitoring program for workers exposed to Cr(VI) compounds include worker education, a preplacement medical examination, and regularly scheduled follow-up medical examinations. Based on the findings from these examinations, more frequent and detailed medical examination may be necessary.

8.6.3.1 Worker education

All workers in the medical monitoring program should be provided with the purposes of the program, the potential health benefits of participation, and program procedures. Workers should be trained in the potential symptoms, findings, and diseases associated with Cr(VI) exposure. They should also be trained in procedures to avoid and minimize their Cr(VI) exposures. They should be instructed to report any accidental exposures to Cr(VI) or incidents involving potentially high exposure levels to their supervisor and the medical director. Workers should inform their health care provider about their workplace exposures and any possible work-related symptoms. They should be instructed to inform their supervisor or the medical director of any symptoms consistent with Cr(VI) exposure if the work-relatedness of the symptoms is confirmed or suspected by a health care provider.

8.6.3.2 Preplacement medical examination

A preplacement examination should be conducted. This examination should consist of a core examination and a respiratory examination. The core examination should be conducted on all workers included in the medical monitoring program. The respiratory examination should be conducted on all workers potentially exposed to Cr(VI) via an airborne route, as specified in Section 8.6.1 above. Routine blood and urine analysis is not recommended, as these tests are of uncertain value as early indicators of potential health effects related to Cr(VI) [NIOSH 2005a].

8.6.3.2.1 Core examination

- A standardized occupational history questionnaire that gathers information on all past jobs, a description of all duties and potential exposures for each job, and a description of all protective equipment the worker has used.

- A detailed medical history including information on conditions such as skin sensitization, occupational asthma, and other dermatologic or respiratory symptoms or disorders that may be exacerbated by exposure to Cr(VI).

- A physical examination of all systems with careful inspection of the skin for evidence of irritation, ulceration, sensitization, or dermatitis, and the mucous membranes, and upper respiratory tract for evidence of irritation, bleeding, ulcerations, or perforation.

8.6.3.2.2 Respiratory examination

- A standardized respiratory symptom questionnaire.

- An evaluation of the worker's ability to use negative or positive pressure respirators.

- A preplacement or baseline spirometric test within 3 months of assignment. Anyone administering spirometric testing as part of the medical monitoring program should have completed a NIOSH-approved training course in spirometry or other equivalent training.

- A baseline chest radiograph within 3 months of assignment.

8.6.3.3 Follow-up medical examinations

All workers in the medical monitoring program should be provided with follow-up medical examinations conducted by a physician or other qualified health care provider. The following recommendations are suggested for workers in good health. Any worker with adverse health effects potentially associated with Cr(VI) exposure should be examined immediately and may require more frequent monitoring and extensive testing.

8.6.3.3.1 Follow-up examination frequency

Each worker should have a thorough medical evaluation of the skin and respiratory tract (upper and lower) conducted every 6 months for the first 2 years of employment and annually thereafter, unless adverse health effects warrant more frequent monitoring. The annual medical examination should be conducted with emphasis on the skin and respiratory tract, and an occupational history update questionnaire should be completed annually.

8.6.3.3.2 Follow-up respiratory examination

The follow-up respiratory examinations, conducted as noted above, should include the following. The respiratory symptom questionnaire should be updated at the time of examination. Spirometric testing should be conducted annually for the first 3 years and every 2 to 3 years thereafter, or as indicated by current medical recommendations and the scientific literature. Based on the findings from these examinations, more frequent and detailed medical examination or testing may be necessary. Interpretation of annual lung function changes within an individual worker are specified and updated by professional organizations, such as the American Thoracic Society (ATS) and the American College of Occupational and Environmental Medicine (ACOEM) [ATS 1995; ACOEM 2011]. Spirometry resources for employers and workers are available on the NIOSH Spirometry Safety and Health Topic Page (http://www.cdc.gov/niosh/topics/spirometry/) [NIOSH 2011].

The value of periodic chest radiographs in a medical surveillance program should be evaluated by a qualified health-care professional, in consultation with the worker, based on current medical recommendations and the scientific literature to assess whether the benefits of testing warrant the additional exposure to radiation. If the qualified health-care professional deems periodic chest radiographs useful, their timing and frequency should take into account the observed latency and natural history of occupational lung cancer associated with Cr(VI), symptoms reported by the worker, and other relevant findings.

Any worker may require more frequent and/or more detailed medical evaluation if he or she has any of the following indications:

- A history of exposure to Cr(VI) compounds, asbestos, any other lung carcinogen, or other respiratory hazard.
- A past or present history of smoking.
- New or worsening dermatologic or respiratory symptoms.
- Other medically significant reason(s) for more detailed assessment.

8.6.4 Medical Reporting

Following each medical examination, the physician or other qualified health-care provider should provide each worker with a written report containing the following:

- The results of any medical tests performed on the worker.
- A medical opinion in plain language about any medical condition that would increase the worker's risk of impairment from exposure to Cr(VI) compounds.
- Recommendations for limiting the worker's exposure to Cr(VI) compounds including the use of appropriate respiratory protective devices or protective clothing.
- Recommendations for further evaluation and treatment of medical conditions detected.

Following each medical examination, the physician should provide a written report to the employer that contains the following:

- Occupationally pertinent results of the medical evaluation.
- A medical opinion about any medical condition that would increase the worker's risk of illness or disease as a result of exposure to Cr(VI) compounds.
- Recommendations for limiting the worker's exposure to Cr(VI) compounds that may include the use of appropriate respiratory protective devices or protective clothing or reassignment to another job, as warranted.
- A statement that the worker has been informed about the results of the medical examination and about medical condition(s) that should have further evaluation or treatment.

Specific findings, test results, or diagnoses that have no bearing on the worker's ability to work with Cr(VI) compounds should not be included in the report to the employer. Safeguards to protect the confidentiality of the worker's medical records should be enforced in accordance with all applicable regulations and guidelines.

8.6.5 Employer Actions

The employer should ensure that the qualified health care provider's recommended restriction of a worker's exposure to Cr(VI) compounds or other workplace hazards is followed, and that the REL for Cr(VI) compounds is not exceeded without requiring the use of PPE. Efforts to encourage worker participation in the medical monitoring program and to report any symptoms promptly to the program director are important to the program's success. Medical evaluations performed as part of the medical monitoring program should be provided by the employer at no cost to the participating workers. Where medical removal or job reassignment is indicated, the affected worker should not suffer loss of wages, benefits, or seniority.

The employer should ensure that the program director regularly collaborates with the employer's safety and health personnel (e.g. industrial hygienists) to identify and control work exposure and activities that pose a risk of adverse health effects.

8.7 Smoking Cessation

Smoking should be prohibited in all areas of any workplaces in which workers are exposed to Cr(VI) compounds. As cigarette smoking is an important cause of lung cancer, NIOSH recommends that smoking be prohibited in the workplace and all workers who smoke participate in a smoking cessation program. Employers are urged to establish smoking cessation programs that inform workers about the hazards of cigarette smoking and provide assistance and encouragement for workers who want to quit smoking. These programs should be offered at no cost to the participants. Information about the carcinogenic effects of smoking should be disseminated. Activities promoting physical fitness and other health lifestyle practices that affect respiratory and overall health should be encouraged through training, employee assistance programs, and/or health education campaigns.

8.8 Record Keeping

Employers should keep employee records on exposure and medical monitoring according to the requirements of 29 CFR 1910.20(d), Preservation of Records.

Accurate records of all sampling and analysis of airborne Cr(VI) conducted in a workplace should be maintained by the employer for at least 30 years. These records should include the name of the worker being monitored; Social Security number; duties performed and job locations; dates and times of measurements; sampling and analytical methods used; type of personal protection used; and number, duration, and results of samples taken.

Accurate records of all medical monitoring conducted in a workplace should be maintained by the employer for 30 years beyond the employee's termination of employment.

Table 8–1. Respiratory protection recommendations for exposure to Cr(VI)* compounds

Airborne Cr(VI) concentration or condition of use†	Minimum respiratory protection
≤ 2 µg/m³ (10 x REL)	Any half mask particulate air-purifying respirator with a N100‡, R100, or P100 filter worn in combination with eye protection.
	If chromyl chloride is present, any half mask air-purifying respirator with canisters providing organic vapor (OV) and acid gas (AG) protection with a N100, R100, or P100 filter worn in combination with eye protection.
≤ 5 µg/m³ (25 x REL)	Any supplied-air respirator with loose-fitting hood or helmet operated in a continuous-flow mode; any powered air-purifying respirator (PAPR) with a HEPA filter with loose-fitting hood or helmet.
	If chromyl chloride is present, any PAPR providing OV and AG protection with a HEPA filter with loose-fitting hood or helmet.
≤ 10 µg/m³ (50 x REL)	Any full facepiece particulate air-purifying respirator with a N100, R100, or P100 filter; any PAPR with full facepiece and HEPA filter; any full facepiece supplied-air respirator operated in a continuous-flow mode.
	If chromyl chloride is present, any full facepiece air-purifying respirator providing OV and AG protection with a N100, R100, or P100 filter; any full facepiece PAPR providing OV and AG protection and a HEPA filter.
≤ 400 µg/m³ (2,000 x REL)	Any supplied-air, pressure-demand respirator with full facepiece.
≤ 2000 µg/m³ (10,000 x REL)	Any self-contained breathing apparatus that is operated in a pressure-demand or other positive-pressure mode, or any supplied-air respirator with a full facepiece that is operated in a pressure-demand or other positive-pressure mode in combination with an auxiliary self-contained positive-pressure breathing apparatus.
> 2000 µg/m³ §	Any self-contained breathing apparatus that has a full facepiece and is operated in a pressure-demand or other positive-pressure mode.
Escape only	Any air-purifying, full-facepiece respirator with a N100, R100, or P100 filter or any appropriate escape-type, self-contained breathing apparatus.
	If chromyl chloride is present, any full facepiece gas mask (14G) with a canister providing OV and AG protection with a N100, R100, or P100 filter or any appropriate escape-type, self-contained breathing apparatus.

*Abbreviations: AG= acid gas; APF = assigned protection factor; Cr(VI) = hexavalent chromium; HEPA = high efficiency particulate air; IDLH = immediately dangerous to life or health; OV=organic vapor; PAPR = powered air-purifying respirator.
†The protection offered by a given respirator is contingent upon (1) the respirator user adhering to complete program requirements (such as those required by OSHA in 29 CFR 1910.134), (2) the use of NIOSH-certified respirators in their approved configuration, and (3) individual fit testing to rule out those respirators with tight-fitting facepieces that cannot achieve a good fit on individual workers.
‡N-100 series particulate filters should not be used in environments where there is potential for exposure to oil mists.
§Immediately Dangerous to Life or Health (IDLH) value for chromic acid and chromates = 15 mg/m³.

References

71 FR 10099 [2006]. Occupational Safety and Health Administration: Occupational exposure to hexavalent chromium; final rule (29 CFR Parts 1910, 1915, 1917, 1918, 1926). Docket No. H-0054A.

76 FR 25569 [2011]. Department of Defense: Defense federal acquisition regulation supplement; minimizing the use of materials containing hexavalent chromium; final rule (48 CFR Parts 233, 252).

ACGIH [2010]. Industrial ventilation: a manual of recommended practice. 27th ed. Cincinnati, OH: American Conference of Governmental Industrial Hygienists, Publication No. 2093.

ACGIH [2011a]. 2011 TLVs and BEIs. Cincinnati, OH: American Conference of Governmental Industrial Hygienists, Publication No. 0111.

ACGIH [20011b]. 2011 Guide to occupational exposure values. Cincinnati, OH: American Conference of Governmental Industrial Hygienists, Publication No. 0389.

ACOEM [2011]. Spirometry in the occupational health setting—2011 update. Townsend MC (lead). American College of Environmental and Occupational Medicine Position Statement [http://www.acoem.org/uploadedFiles/Public_Affairs/Policies_And_Position_Statements/ACOEM%20Spirometry%20Statement.pdf].

Adachi S [1987]. Effect of chromium compounds on the respiratory system: Part 5. Long term inhalation of chromic acid mist in electroplating by C57BL female mice and recapitulation on our experimental studies. Jpn J Ind Health 29(1):17–33.

Adachi S, Yoshimura H, Katayama H, Takemoto K [1986]. Effects of chromium compounds on the respiratory system: Part 4. Long term inhalation of chromic acid mist in electroplating to ICR female mice. Jpn J Ind Health 28(4):283–287.

Ahrenholz SH, Anderson KE [1981]. Health hazard evaluation report: Valley Chrome Platers, Bay City, Michigan. Cincinnati, OH: U.S. Department of Health, Education, and Welfare, Public Health Service, Centers for Disease Control, National Institute for Occupational Safety and Health, NIOSH Report No. HETA-81-085-889.

AIHA [2006]. A strategy for assessing and managing occupational exposures. Ignacio JS, Bullock WH, AIHA, eds. Fairfax, VA; American Industrial Hygiene Association.

AIHA [2011]. The occupational environment: Its evaluation, control, and management. 3rd eds. Anna DH ed. Fairfax, VA; American Industrial Hygiene Association.

Aiyar J, Berkovits HJ, Floyd RA, Wetterhahn KE [1991]. Reaction of chromium(VI) with glutathione or with hydrogen peroxide: identification of reactive intermediates and their role in chromium(VI)-induced DNA damage. Environ Health Perspect 92:53–62.

Alderson MR, Rattan NS, Bidstrup L [1981]. Health of workmen in the chromate-producing industry in Britain. Br J Ind Med 38(2):117–124.

Angerer J, Amin W, Heinrich-Ramm R, Szadkowski D, Lehnert G [1987]. Occupational chronic exposure to metals. Int Arch Occup Environ Health 59(5):503–512.

ANSI [1989]. Practice for occupational and educational personal eye and face protective devices (ANSI Z87.1-1989). Washington, DC: American National Standards Institute.

Ashley K, Howe AM, Demange M, Nygren O [2003]. Sampling and analysis considerations for the determination of hexavalent chromium in workplace air. J Environ Monit 5:707–716.

Ashley K, Applegate GT, Marcy AD, Drake PL, Pierce PA, Carabin N, Demange M [2009]. Evaluation of sequential extraction procedures for soluble and insoluble hexavalent chromium compounds in workplace air samples. J Environ Monit 11(2):318–325.

ASTM [2002]. Standard test method for the determination of hexavalent chromium in workplace air by ion chromatography and spectrophotometric measurement using 1,5-diphenylcarbazide. West Conshohocken, PA: ASTM International, Publication No. ASTM D6832-02.

ASTM [2003]. Standard practice for collection of settled dust samples using wipe sampling methods for subsequent determination of metals. West Conshohocken, PA: ASTM International, Publication No. ASTM D6966-03.

ASTM [2011]. Selected technical papers: Surface and dermal sampling. West Conshohocken, PA: ASTM International, Publication No. ASTM STP1533-EB.

ATS [1995]. Standardization of spirometry. Am J Respir Crit Care Med *152*:1107–1136.

ATSDR [2012]. Toxicological profile for chromium. Atlanta, GA: U.S. Department of Health and Human Services, Public Health Service, Agency for Toxic Substances and Disease Registry.

Avnstorp C [1989]. Prevalence of cement eczema in Denmark before and since addition of ferrous sulfate to Danish cement. Acta Derm Venereol *69*(2):151–155.

Axelsson G, Rylander R, Schmidt A [1980]. Mortality and incidence of tumours among ferrochromium workers. Br J Ind Med *37*(2):121–127.

Baetjer AM [1950]. Pulmonary carcinoma in chromate workers II. Incidence on basis of hospital records. Arch Ind Hyg Occup Med *2*(5):505–516.

Bagdon RE, Hazen RE [1991]. Skin permeation and cutaneous hypersensitivity as a basis for making risk assessments of chromium as a soil contaminant. Environ Health Perspect *92*:111–119.

Barnowska-Dutkiewica B [1981]. Absorption of hexavalent chromium by skin in man. Arch Toxicol *47*(1):47–50.

Battista G, Comba P, Orsi D [1995]. Nasal cancer in leather workers: an occupational disease. J Cancer Res Clin Oncol *121*(1):1–6.

Becker N [1999]. Cancer mortality among arc welders exposed to fumes containing chromium and nickel. Results of a third follow-up: 1989–1995. J Occup Environ Med *41*(4):294–303.

Becker N, Claude J, Frentzel-Beyme R [1985]. Cancer risk of arc welders exposed to fumes containing chromium and nickel. Scand J Work Environ Health *11*(2):75–82.

Berlinger B, Ellingsen DG, Náray M, Zaray G, Thomassen Y [2008]. A study of the bio-accessibility of welding fumes. J Environ Monit *10*(12):1448–1453.

Bertazzi PA, Zocchetti C, Terzaghi GF, Riboldi L, Guercilena S, Beretta F [1981]. Mortality experience of paint production workers. Med Lav *6*:465–472.

Bidstrup PL, Case RAM [1956]. Carcinoma of the lung in workmen in the bichromates-producing industry in Great Britain. Br J Ind Med *13*(4):260–264.

Birk T, Mundt KA, Dell LD, Luippold RS, Miksche L, Steinmann-Steiner-Haldenstaett W, Mundt DJ [2006]. Lung cancer mortality in the German chromate industry, 1958–1998. J Occup Environ Med *48*(4):426–433.

Blade LM, Yencken MS, Wallace ME, Catalano JD, Khan A, Topmiller JL, Shulman SA, Martinez A, Crouch KG, Bennett JS [2007]. Hexavalent chromium exposures and exposure-control technologies in American enterprise: results of a NIOSH field research study. J Occup Environ Hyg *4*(8):596–618.

Blair A [1980]. Mortality among workers in the metal polishing and plating industry, 1951–1969. J Occup Med *22*(3):158–162.

Blair A, Mason TJ [1980]. Cancer mortality in United States countries with metal electroplating industries. Arch Environ Health *35*(2):92–94.

Blot WJ, Fryzek JP, Henderson BE, Sadler CJ, McLaughlin JK [2000]. A cohort mortality study among gas generator utility workers. J Occup Environ Med *42*(2):194–199.

Boiano JM, Wallace ME, Sieber WK, Groff JH, Wang J, Ashley KE [2000]. Comparison of three sampling and analytical methods for the determination of airborne hexavalent chromium. J Environ Monit *2*(4):329–333.

Bonde JPE, Olsen JH, Hansen KS [1992]. Adverse pregnancy outcome and childhood malignancy with reference to paternal welding exposure. Scand J Work Environ Health *18*(3):169–177.

Bourne HG Jr., Yee HT [1950]. Occupational cancer in a chromate plant. An environmental appraisal. Ind Med Surg *19*(12):563–567.

Brand P, Gube M, Gerards K, Bertram J, Kaminski H, John AC, Kuhlbusch T, Wiemann M, Eisenbeis C, Winkler R, Kraus T [2010]. Internal exposure, effect monitoring, and lung function in welders after acute short-term exposure to welding fumes from different welding processes. J Occup Environ Med *52*(9):887–892.

Braver ER, Infante P, Chu K [1985]. An analysis of lung cancer risk from exposure to hexavalent chromium. Teratog Carcinog Mutagen *5*(5):365–378.

Bright P, Burge PS, O'Hickey SP, Gannon PFG, Robertson AS, Boran A [1997]. Occupational asthma due to chrome and nickel electroplating. Thorax 52(1):28–32.

Brinton HP, Frasier ES, Koven AL [1952]. Morbidity and mortality experience among chromate workers. Public Health Reports 67(9):835–847.

Buckell M, Harvey DG [1951]. An environmental study of the chromate industry. Br J Ind Med 8(4):298–301.

Budanova L [1980]. Clinical symptoms and dynamics of occupational bronchial asthma induced by exposure to hexavalent chromium in alumina industry workers. Gig Tr Prof Zabol 10:43–46.

Burges DCL, Gannon PFG, Boran A, Burge PS [1994]. Occupational asthma in hard chrome electroplaters. In: Proceedings of the 9th international symposium on epidemiology in occupational health, September 23–25, 1992, Cincinnati, OH. Cincinnati, OH: US Department of Health and Human Services, Public Health Service, National Institute for Occupational Safety and Health, DHHS (NIOSH) Publication No. 94-112, pp. 476–481.

Burkhart JE, Knutti EB [1994]. Health hazard evaluation and technical assistance report: Dee Zee Manufacturing, Des Moines, Iowa. Cincinnati, OH: U.S. Department of Health and Human Services, Centers for Disease Control, National Institute for Occupational Safety and Health, NIOSH Report No. HETA 91-0142-2434.

Burrows D [1983]. Adverse chromate reactions on the skin. In: Burrows D, ed. Chromium: metabolism and toxicity. Boca Raton, FL: CRC Press, pp.137–163.

Burrows D [1987]. Chromate dermatitis. In: Maibach HI, ed. Occupational and industrial dermatology. 2nd ed. Chicago, IL: Year Book Medical Publishers, Inc., pp. 406–420.

Burrows D, Adams RM, Flint GN [1999]. Metals. In: Adams RM, ed. Occupational skin disease. 3rd ed. Philadelphia, PA: W.B. Saunders Company, pp. 395–433.

Centaur Associates, Inc [1981]. Technological and economic analysis of regulating occupational exposure to chromium. Prepared for the Occupational Safety and Health Administration.

CFR. Code of Federal regulations. Washington, DC: U.S. Government Printing Office, Office of the Federal Register.

Chan-Yeung M [1995]. Occupational asthma. Environ Health Perspect 103(Suppl 6):249–252.

Chen C-J, Shih T-S, Chang H-Y, Yu H-S, Wu J-D, Sheu S-C, Wu C-E, Chou T-C [2008]. The total body burden of chromium associated with skin disease and smoking among cement workers. Sci Tot Environ 391(1):76–81.

Chen F, Castranova V, Shi X, Demers LM [1999]. New insights into the role of nuclear factor- κB, a ubiquitous transcription factor in the initiation of disease. Clin Chem 45(1):7–17.

Chen F, Ding M, Lu Y, Leonard SS, Vallyathan V, Castranova V, Shi X [2000]. Participation of MAP kinase p38 and IκB kinase in chromium (VI)-induced NF-κB and AP-1 activation. J Environ Pathol Toxicol Oncol 19(3):231–238.

Chromate Toxicity Review Committee [2001]. Scientific review of toxicological and human health issues related to the development of a public health goal for chromium(VI). Report prepared by the Chromate Toxicity Review Committee, August 31, 2001. [http://www.oehha.ca.gov/public_info/facts/pdf/CrPanelRptFinal901.pdf]. Date accessed: June 18, 2003.

Cohen MD, Kargacin B, Klein CB, Costa M [1993]. Mechanisms of chromium carcinogenicity and toxicity. Crit Rev Toxicol 23(3):255–281.

Cole P, Rodu B [2005]. Epidemiologic studies of chrome and cancer mortality: a series of meta-analyses. Reg Toxicol Pharmacol 43(3):225–231.

Comba P, Barbieri PG, Battista G, Belli S, Ponterio F, Zanetti D, Axelson O [1992]. Cancer of the nose and paranasal sinuses in the metal industry: a case-control study. Br J Ind Med 49(3):193–196.

Committee on Biologic Effects of Atmospheric Pollutants [1974]. In: Chromium. Washington, DC: National Academy of Sciences.

Corbett GE, Dodge DG, O'Flaherty E, Liang J, Throop L, Finley BL, Kerger BD [1998]. In vitro reduction kinetics of hexavalent chromium in human blood. Environ Res 78(1):7–11.

Corbett GE, Finley BL, Paustenbach DJ, Kerger BD [1997]. Systemic uptake of chromium in human volunteers following dermal contact with hexavalent chromium (22 mg/L). J Expo Anal Environ Epidemiol 7(2):179–189.

Costa M [1997]. Toxicity and carcinogenicity of Cr(VI) in animal models and humans. Crit Rev Toxicol 27(5):431–442.

CPWR [1999]. An employer's guide to skin protection. FOF Communications. [http://www.elcosh.org/en/document/465/d000457/an-employers-guide-to-skin-protection.html]. Date accessed: November 16, 2004.

CRIOS [2003]. Chromium and chromium compounds: toxicology. Carcinogenic risk in occupational settings (CRIOS). [http://www.crios.be/Chromium/toxicology.htm]. Date accessed: April 1, 2003.

Criteria Group for Occupational Standards [2000]. Scientific basis for Swedish occupational standards XXI. Consensus report for chromium and chromium compounds. Arbete och Hälsa Part 22:18–40.

Cross HJ, Faux SP, Sadhra S, Sorahan T, Levy LS, Aw TC, Braithwaite R, McRoy C, Hamilton L, Calvert I [1997]. Criteria document for hexavalent chromium. Paris, France: International Chromium Development Association.

Crump C, Crump K, Hack E, Luippold R, Mundt K, Liebig E, Panko J, Paustenbach D, Proctor D [2003]. Dose-response and risk assessment of airborne hexavalent chromium and lung cancer mortality. Risk Anal 23(6):1147–1163.

Dalager NA, Mason TJ, Fraumeni JF Jr., Hoover R, Payne WW [1980]. Cancer mortality among workers exposed to zinc chromate paints. J Occup Med 22(1):25–29.

Dankovic D, Kuempel E, Wheeler M [2007]. An approach to risk assessment for TiO_2. Inhal Toxicol 19(Suppl 1):205–212.

Davies JM [1978]. Lung cancer mortality of workers making chrome pigments [letter]. Lancet 1(8060):384.

Davies JM [1979]. Lung cancer mortality of workers in chromate pigment manufacture: an epidemiological survey. J Oil Col Chem Assoc 62:157–163.

Davies JM [1984]. Lung cancer mortality among workers making lead chromate and zinc chromate pigments at three English factories. Br J Ind Med 41(2):158–169.

Davies JM, Easton DF, Bidstrup PL [1991]. Mortality from respiratory cancer and other causes in United Kingdom chromate production workers. Br J Ind Med 48(5):299–313.

Dayan AD, Paine AJ [2001]. Mechanism of chromium toxicity, carcinogenicity and allergenicity: review of literature from 1985 to 2000. Hum Exp Toxicol 20:439–451.

De Flora S, Bagnasco M, Serra D, Zanacchi P [1990]. Genotoxicity of chromium compounds. A review. Mutation Res 238(2):99–172.

De Flora S, Wetterhahn KE [1989]. Mechanisms of chromium metabolism and genotoxicity. Life Chem Rep 7:169–244.

De Marco R, Bernardinelli L, Mangione MP [1988]. Death risk due to cancer of the respiratory apparatus in chromate production workers. Med Lav 79(5):368–376.

Deschamps F, Moulin JJ, Wild P, Labriffe H, Haguenoer JM [1995]. Mortality study among workers producing chromate pigments in France. Int Arch Occup Environ Health 67(3):147–152.

Ding M, Shi X [2002]. Molecular mechanisms of Cr(VI)-induced carcinogenesis. Mol Cell Biochem 234–235(1-2):293–300.

Douglas GR, Bell RD, Grant CE, Wytsma JM, Bora KC [1980]. Effect of lead chromate on chromosome aberration, sister-chromatid exchange and DNA damage in mammalian cells in vitro. Mutat Res 77(2):157–163.

DECOS (Dutch Expert Committee on Occupational Standards) [1998]. Chromium and its inorganic compounds. Health-based recommended occupational exposure limit (revised version). Health Council of the Netherlands (Gezondheidsraad) No. 1998/01(R)WGD.

Electric Power Research Institute [2009]. Airborne hexavalent chromium during welding and thermal metal cutting. Report No. 1019015. Palo Alto, CA: Electric Power Research Institute.

El Ghissassi F, Baan R, Straif K, Grosse Y, Secretan B, Bouvard V, Benbrahim-Tallaa L, Guha N, Freeman C, Galichet L, Cogliano V [2009]. A review of human carcinogens—part D: radiation. Lancet Oncol 10(8):751–752.

Enterline PE [1974]. Respiratory cancer among chromate workers. J Occup Med 16(8):523–526.

EPA [1984]. Health assessment document for chromium. Research Triangle Park, NC: Environmental Assessment and Criteria Office, U.S. Environmental Protection Agency. EPA 600/4-79-020.

EPA [1998]. Toxicological review of hexavalent chromium (CAS No. 18540-29-9) in support of summary information on the Integrated Risk Information System (IRIS). Washington, DC: U.S. Environmental Protection Agency [http://www.epa.gov/iris].

EPA [1999]. Integrated risk information system: Chromium (VI) (CASRN 18540-29-9). Washington, DC: U.S. Environmental Protection Agency [http://www.epa.gov/iris/subst/0144.htm].

European Parliament and the Council of the European Union [2003]. Directive 2003/53/EC of the European Parliament and of the Council of 18 June 2003 amending for the 26th time Council Directive 76/769/EEC relating to restrictions on the marketing and use of certain dangerous substances and preparations (nonylphenol, nonylphenol ethoxylate and cement). OJEU 178:24–27.

Exponent for Chromium Coalition [2002]. Critique of two studies by Gibb et al.: Lung cancer among workers in chromium chemical production; Clinical findings of irritation among chromium chemical production workers. Prepared by Exponent for Chromium Coalition. Doc. No. 8601838.001 0101 0602 DP06.

Fairchild EJ [1976]. Guidelines for a NIOSH policy on occupational carcinogenesis. Annals NY Acad Sci 271:200–207.

Federal Security Agency, Public Health Service, Division of Occupational Health of the Bureau of State Services [1953]. Health of workers in chromate producing industry—a study. Public Health Service Publication No. 192. Washington, DC. U.S. Government Printing Office.

Fiore S [2006]. Reducing exposure to hexavalent chromium in welding fumes. Welding Journal 85(8):38–42.

Fornace AJ Jr., Seres JF, Lechner JF, Harris CC [1981]. DNA-protein cross-linking by chromium salts. Chem Biol Interact 36(3):345–354.

Forsberg K, Keith LH [1999]. Chemical protective clothing performance index. 2nd ed. New York: John Wiley and Sons.

Franchini I, Magnani F, Mutti A [1983]. Mortality experience among chromeplating workers. Initial findings. Scand J Work Environ Health 9(3):247–252.

Frentzel-Beyme R [1983]. Lung cancer mortality of workers employed in chromate pigment factories. A multicentric European epidemiological study. J Cancer Res Clin Oncol 105(2):183–188.

Furst A, Schlauder M, Sasmore DP [1976]. Tumorigenic activity of lead chromate. Cancer Res 36(5):1779–1783.

Gad [1986]. Acute toxicity of four chromate salts. In: Serrone DM, ed. Chromium symposium 1986: an update. Pittsburgh, PA: Industrial Health Foundation, pp. 43–58.

Gambelunghe A, Piccinini R, Ambrogi M, Villarini M, Moretti M, Marchetti C, Abbritti G, Muzi G [2003]. Primary DNA damage in chrome-plating workers. Toxicology 188(2–3):187–195.

Gammelgaard B, Fullerton A, Avnstorp C, Menne T [1992]. Permeation of chromium salts through human skin in vitro. Contact Dermatitis 27(5):302–310.

Gao M, Binks SP, Chipman JK, Levy LS, Braithwaite RA, Brown SS [1992]. Induction of DNA strand breaks in peripheral lymphocytes by soluble chromium compounds. Hum Exp Toxicol 11(2):77–82.

Gao M, Levy LS, Faux SP, Aw TC, Braithwaite RA, Brown SS [1994]. Use of molecular epidemiological techniques in a pilot study on workers exposed to chromium. Occup Environ Med 51(10):663–668.

Gao N, Jiang B, Leonard SS, Corum L, Zhang Z, Roberts JR, Antonini J, Zheng JZ, Flynn DC, Castranova V, Shi X [2002]. p38 Signaling-mediated hypoxia-induced factor 1α and vascular endothelial growth factor induction by Cr(VI) in DU145 human prostate carcinoma cells. J Biol Chem 277(47):45041–45048.

Geier J, Krautheim A, Uter W, Lessmann H, Schnuch A [2010]. Occupational contact allergy in the building trade in Germany: influence of preventive measures and changing exposure. Int Arch Occup Environ Health 84(4):403–411.

Gibb HJ, Chen CW, Hiremath CB [1986]. Carcinogen risk assessment of chromium compounds. In: Serrone D, ed. Proceedings of chromium symposium. Pittsburgh, PA: Industrial Health Foundation, pp. 248–309.

Gibb H, Hoffman HJ, Haver C [2011]. Biologic implications from an epidemiologic study of chromate production workers. Open Epi Journ 4:54–59.

Gibb HJ, Lees PS, Pinsky PF, Rooney BC [2000a]. Clinical findings of irritation among chromium chemical production workers. Am J Ind Med 38(2):127–131.

Gibb HJ, Lees PSJ, Pinsky PF, Rooney BC [2000b]. Lung cancer among workers in chromium chemical production. Am J Ind Med 38(2):115–126.

Glaser U, Hochrainer D, Kloppel H, Kuhnen H [1985]. Low level chromium(VI) inhalation effects on alveolar macrophages and immune functions in Wistar rats. Arch Toxicol 57(4):250–256.

Glaser U, Hochrainer D, Kloppel H, Oldiges H [1986]. Carcinogenicity of sodium dichromate and chromium (VI/III) oxide aerosols inhaled by male Wistar rats. Toxicology 42(2–3):219–232.

Glaser U, Hochrainer D, Oldiges H [1988]. Investigations of the lung carcinogenic potentials of sodium dichromate and Cr VI/III oxide aerosols in Wistar rats. Environ Hyg 1:111–116.

Glaser U, Hochrainer D, Steinhoff D [1990]. Investigation of irritating properties of inhaled CrVI with possible influence on its carcinogenic action. Environ Hyg 2:235–245.

Goh CL, Gan SC [1996]. Change in cement manufacturing process, a cause for decline in chromate allergy? Contact Dermatitis 34(1):51–54.

Goldbohm RA, Tielemans ELFP, Heederik D, Rubingh CM, Dekkers S, Willems MI, Dinant Kroese E [2006]. Risk estimation for carcinogens based on epidemiological data: a structure approach, illustrated by an example on chromium. Regul Toxicol Pharmacol 44(3):294–210.

Government of Canada, Environment Canada, Health Canada [1994]. Chromium and its compounds. Canadian Environmental Protection Act, Priority Substances List Assessment Report. Catalogue No. En 40-215/39E. [http://www.ec.gc.ca/substances/ese/eng/psap/PSL1_reports/chromium.pdf] Date accessed: June 20, 2003.

Gray SJ, Sterling K [1950]. The tagging of red blood cells and plasma proteins with radioactive chromium. J Clin Invest 29(12):1604–1613.

Gylseth B, Gundersen N, Langård S [1977]. Evaluation of chromium exposure based on a simplified method for urinary chromium determination. Scand J Work Environ Health 3(1):28–31.

Haguenoer JM, Dubois G, Frimat P, Cantineau A, Lefrançois H, Furon D [1981]. Mortality due to bronchopulmonary cancer in a factory producing pigments based on lead and zinc chromates [French]. In: Prevention of occupational cancer, International Symposium (Occupational Safety and Health Series), Geneva, International Labour Office (No. 46), pp. 168–176.

Haines AT, Nieboer E [1988]. Chromium hypersensitivity. In: Nriagu JO, Nieboer E, eds. Chromium in the natural and human environments. New York, NY: John Wiley and Sons, pp. 497–532.

Handley J, Burrows D [1994]. Dermatitis from hexavalent chromate in the accelerator of an epoxy sealant (PR1422) used in the aircraft industry. Contact Dermatitis 30(4):193–196.

Hayes RB [1980]. Cancer and occupational exposure to chromium chemicals. Rev Cancer Epidemiol 1:293–333.

Hayes RB [1988]. Review of occupational epidemiology of chromium chemicals and respiratory cancer. Sci Total Environ 71(3):331–339.

Hayes RB [1997]. The carcinogenicity of metals in humans. Cancer Causes Control 8(3):371–385.

Hayes RB, Lilienfeld AM, Snell LM [1979]. Mortality in chromium chemical production workers: a prospective study. Int J Epidemiology 8(4):365–374.

Hayes RB, Sheffet A, Spirtas R [1989]. Cancer mortality among a cohort of chromium pigment workers. Am J Ind Med 16(2):127–133.

Hazelwood KJ, Drake PL, Ashley K, Marcy D [2004]. Field method for the determination of insoluble or total hexavalent chromium in workplace air. J Occup Environ Hyg 1(9):613–619.

Health Council of the Netherlands: Committee for Compounds Toxic to Reproduction [2001]. Chromium VI and its compounds: Evaluation of the effects on reproduction, recommendation for classification. The Hague: Health Council of the Netherlands. Pub No. 2001/01OSH.

Heung W, Yun MJ, Chang DP, Green PG, Halm C [2007]. Emissions of chromium (VI) from arc welding. J Air Waste Manag Assoc 57(2):252–260.

Hills L, Johansen VC [2007]. Hexavalent chromium in cement manufacturing: literature review. Skokie, IL: Portland Cement Association, PCA R&D Serial No. 2983.

Hjollund NHI, Bonde JPE, Hansen KS [1995]. Male-mediated spontaneous abortion with reference to stainless steel welding. Scand J Work Environ Health 21:272–276.

Hjollund NHI, Bonde JPE, Jensen TK, Henriksen TB, Kolstad HA, Ernst E, Giwercman A, Skakkebaek NE, Olsen J [1998]. A follow-up study of male exposure to welding and time to pregnancy. Reprod Toxicol 12(1):29–37.

Hjollund NHI, Bonde JPE, Jensen TK, Henriksen TB, Andersson A-M, Kolstad HA, Ernst E, Giwercman A, Skakkebaek NE, Olsen J [2000]. Male-mediated spontaneous abortion among spouses of stainless steel welders. Scand J Work Environ Health 26(3):187–192.

Holmes AL, Wise SS, Sandwick SJ, Lingle WL, Negron VC, Thompson WD, Wise JP Sr. [2006a]. Chronic exposure to lead chromate causes centrosome abnormalities and aneuploidy in human lung cells. Cancer Res 66(8):4041–4048.

Holmes AL, Wise SS, Sandwick SJ, Wise JP Sr. [2006b]. The clastogenic effects of chronic exposure to particulate and soluble Cr(VI) in human lung cells. Mut Res 610(1–2):8–13.

Holmes AL, Wise SS, Wise JP Sr. [2008]. Carcinogenicity of hexavalent chromium. Indian J Med Res 128(4):353–372.

Hueper WC [1961]. Environmental carcinogenesis and cancers. Cancer Res 21:842–857.

Hughes K, Meek ME, Seed LJ, Shedden J [1994]. Chromium and its compounds: evaluation of risks to health from environmental exposure in Canada. J Environ Sci Health Part C Environ Carcino & Ecotox Revs 12(2):237–255.

Huvinen M, Uitti J, Zitting A, Roto P, Virkola K, Kuikka P, Laippala P, Aitio A [1996]. Respiratory health of workers exposed to low levels of chromium in stainless steel production. Occup Environ Med 53(11):741–747.

Huvinen M, Uitti J, Oksa P, Palmroos P, Laippala P [2002a]. Respiratory health effects of long-term exposure to different chromium species in stainless steel production. Occup Med (Lond) 52(4):203–212.

Huvinen M, Mäkitie A, Järventaus H, Wolff H, Stjernvall T, Hovi A, Hirvonen A, Ranta R, Nurminen M, Norppa H [2002b]. Nasal cell micronuclei, cytology and clinical symptoms in stainless steel production workers exposed to chromium. Mutagenesis 17(5):425–429.

IARC [1990]. IARC monographs on the evaluation of the carcinogenic risk of chemicals to man: chromium, nickel, and welding. Vol. 49. Lyon, France: World Health Organization, International Agency for Research on Cancer, pp. 49–256.

IARC [2012]. IARC monographs on the evaluation of the carcinogenic risks to humans: a review of human carcinogens: arsenic, metals, fibres, and dusts. Vol. 100C. Lyon, France: World Health Organization, International Agency for Research on Cancer, pp. 147–168.

ICDA (International Chromium Development Association) [1997]. Criteria document for hexavalent chromium. Paris, France.

IPCS (WHO/International Programme on Chemical Safety) [1994]. International Chemical Safety card: Chromium oxide [http://www.cdc.gov/niosh/ipcsneng/neng1194.html]. Date accessed: July 21, 2004.

ISO [2001]. ISO 10882-1: Health and safety in welding and allied processes—sampling of airborne particles and gases in the operator's breathing zone—Part 1: Sampling of airborne particles. Geneva, Switzerland: International Organization for Standardization.

ISO [2005]. ISO 16740: Workplace air—determination of hexavalent chromium in airborne particulate matter—method by ion chromatography and spectrophotometric measurement using diphenyl carbazide. Geneva, Switzerland: International Organization for Standardization.

Itoh T, Takahashi K, Okubo T [1996]. Mortality of chromium plating workers in Japan—a 16-year follow-up study. J UOEH. 18(1):7–18.

Järvholm B, Thiringer G, Axelson O [1982]. Cancer morbidity among polishers. Br J Ind Med 39(2):196–197.

Jennette KW [1979]. Chromate metabolism in liver microsomes. Biol Trace Elem Res 1:55–62.

Ji L, Arcinas M, Boxer LM [1994]. NF-κB sites function as positive regulators of expression of the translocated c-myc allele in Burkitt's Lymphoma. Mol Cell Biol 14(12):7967–7974.

Johansen M, Overgaard E, Toft A [1994]. Severe chronic inflammation of the mucous membranes in the eyes and upper respiratory tract due to work-related exposure to hexavalent chromium. J Laryngol Otol 108(7):591–592.

Kano K, Horikawa M, Utsunomiya T, Tati M, Satoh K, Yamaguchi S [1993]. Lung cancer mortality among a cohort of male chromate pigment workers in Japan. Int J Epidemiol 22(1):16–22.

Kasprzak KS [1991]. The role of oxidative damage in metal carcinogenicity. Chem Res Toxicol 4(6):604–615.

Katz SA, Salem H [1993]. The toxicology of chromium with respect to its chemical speciation: a review. J Appl Toxicol 13(3):217–224.

Keane M, Stone S, Chen B, Slaven J, Schwegler-Berry D, Antonini J [2009]. Hexavalent chromium content in stainless steel welding fumes is dependent on the welding process and shield gas type. J Environ Monit 11(2):418–424.

Keskinen H, Kalliomäki P-L, Alanko K [1980]. Occupational asthma due to stainless steel welding fumes. Clin Allergy 10(2):151–159.

Klein CB, Frenkel K, Costa M [1991]. The role of oxidative processes in metal carcinogenesis. Chem Res Toxicol 4(6):592–604.

Korallus U, Lange HJ, Neiss A, Wüstefeld E, Zwingers T [1982]. Relationship between prevention measures and mortality due to bronchial cancer in the chromate industry. Arb Sozialmed Praven 17(7):159–167.

Korallus U, Ulm K, Steinmann-Steiner-Haldenstaett W [1993]. Bronchial carcinoma mortality in the German chromate-producing industry: the effects of process modification. Int Arch Occup Environ Health 65(3):171–178.

K.S. Crump Division [1995]. Evaluation of epidemiological data and risk assessment for hexavalent chromium. Prepared for Occupational Safety and Health Administration. Contract No. J-9-F-1-0066 Modification No. 1.

Kuo HW, Chang SF, Wu FY [2003]. Chromium (VI) induced oxidative damage to DNA: increase of urinary 8-hydroxydeoxyguanosine concentrations (8-OHdG) among electroplating workers. Occup Environ Med 60(8):590–594.

Langård S [1983]. The carcinogenicity of chromium compounds in man and animals. In: Burrows D, ed. Chromium: metabolism and toxicity. Boca Raton, FL: CRC Press, pp. 13–30.

Langård S [1990]. One hundred years of chromium and cancer: a review of epidemiological evidence and selected case reports. Am J Ind Med 17(2):189–215.

Langård S [1993]. Role of chemical species and exposure characteristics in cancer among persons occupationally exposed to chromium compounds. Scand J Work Environ Health 19(Suppl 1):81–89.

Langård S, Andersen A, Gylseth B [1980]. Incidence of cancer among ferrochromium and ferrosilicon workers. Br J Ind Med 37(2):114–120.

Langård S, Andersen A, Ravnestad J [1990]. Incidence of cancer among ferrochromium and ferrosilicon workers: an extended observation period. Br J Ind Med 47(1):14–19.

Langård S, Norseth T [1975]. A cohort study of bronchial carcinomas in workers producing chromate pigments. Br J Ind Med 32(1):62–65.

Langård S, Norseth T [1979]. Cancer in the gastrointestinal tract in chromate pigment workers. Arh Hig Rada Toksikol 30(Suppl):301–304.

Langård S, Vigander T [1983]. Occurrence of lung cancer in workers producing chromium pigments. Br J Ind Med 40(1):71–74.

Lee CR, Yoo CI, Lee J, Kang SK [2002]. Nasal septum perforation of welders. Ind Health 40(3):286–289.

Lees PSJ [1991]. Chromium and disease: review of epidemiologic studies with particular reference to etiologic information provided by measures of exposure. Environ Health Perspect 92:93–104.

Leidel NA, Busch KA [1975]. Statistical methods for determining non-compliance. Am Ind Hyg Assoc J 36(11):839–40.

Leonard SS, Roberts JR, Antonini JM, Castranova V, Shi X [2004]. PbCrO4 mediates cellular responses via reactive oxygen species. Mol Cell Biochem 255(1–2):171–179.

Leroyer C, Dewitte JD, Bassanets A, Boutoux M, Daniel C, Clavier J [1998]. Occupational asthma due to chromium. Respiration 65(5):403–405.

Levy LS, Martin PA, Bidstrup PL [1986]. Investigation of the potential carcinogenicity of a range of chromium containing materials on rat lung. Brit J Ind Med 43(4):243–256.

Li H, Chen Q, Li S, Yao W, Li L, Shi X, Wang L, Castranova V, Vallyathan V, Ernst E, Chen C [2001]. Effect of Cr(VI) exposure on sperm quality: human and animal studies. Ann Occup Hyg 45(7):505–511.

Lindberg E, Hedenstierna G [1983]. Chrome plating: symptoms, findings in the upper airways, and effects on lung function. Arch Environ Health 38(6):367–374.

Lindberg E, Vesterberg O [1983]. Monitoring exposure to chromic acid in chromeplating by measuring chromium in urine. Scand J Work Environ Health 9(4):333–340.

Liu CS, Kuo HW, Lai JS, Lin TI [1998]. Urinary N-acetyl-beta-glucosaminidase as an indicator of renal dysfunction in electroplating workers. Int Arch Occup Environ Health 71(5):348–352.

Liu KJ, Mader K, Shi X, Swartz HM [1997a]. Reduction of carcinogenic chromium(VI) on the skin of living rats. Magn Reson Med 38(4):524–526.

Liu KJ, Shi X, Dalal NS [1997b]. Synthesis of Cr(VI)-GSH. Its identification and its free hydroxyl generation: a model compound for Cr(VI) carcinogenicity. Biochem Biophys Res Commun 235(1):54–58.

Luippold RS, Mundt KA, Austin RP, Liebig E, Panko J, Crump C, Crump K, Proctor D [2003]. Lung cancer mortality among chromate production workers. Occup Environ Med 60(6):451–457.

Luippold RS, Mundt KA, Dell LD, Birk T [2005]. Low-level hexavalent chromium exposure and rate of mortality among US chromate production employees. J Occup Environ Med 47(4):381–385.

Luo H, Lu Y, Shi X, Mao Y, Dalal NS [1996]. Chromium (IV)-mediated Fenton-like reaction causes DNA damage: implication to genotoxicity of chromate. Ann Clin Lab Sci 26(2):185–191.

Machle W, Gregorius F [1948]. Cancer of the respiratory system in the United States chromate-producing industry. Pub Health Rep 63(35):1114–1127.

Makinen M, Linnainmaa M [2004a]. Dermal exposure to chromium in electroplating. Ann Occup Hyg 48(3):277–283.

Makinen M, Linnainmaa M [2004b]. Dermal exposure to chromium in the grinding of stainless and acid-proof steel. Ann Occup Hyg 48(3):197–202.

Mali JWH, Van Kooten WJ, Van Neer FCJ [1963]. Some aspects of the behavior of chromium compounds in the skin. J Invest Dermatol 41:111–122.

Malsch PA, Proctor DM, Finley BL [1994]. Estimation of a chromium inhalation reference concentration using the benchmark dose method: a case study. Regul Toxicol Pharmacol 20(1 Pt 1):58–82.

Mancuso TF [1975]. Consideration of chromium as an industrial carcinogen. Presented at the International Conference of Heavy Metals in the Environment, Toronto, Canada.

Mancuso TF [1997]. Chromium as an industrial carcinogen: Part I. Am J Ind Med 31(2):129–139.

Mancuso TF, Hueper WC [1951]. Occupational cancer and other health hazards in a chromate plant: a medical appraisal. I. Lung cancers in chromate workers. Ind Med Surg 20(8):358–363.

Marlow D, Wang J, Wise TJ, Ashley K [2000]. Field test of portable method for the determination of hexavalent chromium in workplace air. Am Lab 32(15):26–28.

Meeker JD, Susi P, Flynn MR [2010]. Hexavalent chromium exposure and control in welding tasks. J Occup Environ Hyg 7(11):607–615.

Merck [2006]. The Merck Index. 14th ed. Whitehouse Station, NJ: Merck & Co. Accessed online March 12, 2007.

Meridian Research [1994]. Selected chapters of an economic impact analysis for a revised OSHA standard for chromium VI: introduction, industry profiles, technological feasibility (for 6 industries) and environmental impacts. Final Report. Contract No. J-0-F-4-0012 Task Order No. 11.

Mikoczy Z, Hagmar L [2005]. Cancer incidence in the Swedish leather tanning industry: updated findings 1958–99. Occup Environ Med 62(7):461–464.

Miksche LW, Lewalter J [1997]. Health surveillance and biological effect monitoring for chromium-exposed workers. Regul Toxicol Pharmacol 26(1 Pt 2):S94–S99.

Milatou-Smith R, Gustavsson A, Sjögren B [1997]. Mortality among welders exposed to high and to low levels of hexavalent chromium and followed for more than 20 years. Int J Occup Environ Health 3(2):128–131.

Minoia C, Cavalleri A [1988]. Chromium in urine, serum and red blood cells in the biological monitoring of workers exposed to different chromium valency states. Sci Total Environ 71(3):323–327.

Montanaro F, Ceppi M, Demers PA, Puntoni R, Bonassi S [1997]. Mortality in a cohort of tannery workers. Occup Environ Med 54(8):588–591.

Moulin JJ, Wild P, Mantout B, Fournier-Betz M, Mur JM, Smagghe G [1993]. Mortality from lung cancer and cardiovascular diseases among stainless-steel producing workers. Cancer Causes Control 4(2):75–81.

Moulin JJ, Clavel T, Roy D, Dananché B, Marquis N, Févotte J, Fontana JM [2000]. Risk of lung cancer in workers producing stainless steel and metallic alloys. Int Arch Occup Environ Health 73(3):171–180.

Nethercott J, Paustenbach D, Adams R, Fowler J, Marks J, Morton C, Taylor J, Horowitz S, Finley B [1994]. A study of chromium induced allergic contact dermatitis with 54 volunteers: implications for environmental risk assessment. Occup Environ Med 51(6):371–380.

Nethercott J, Paustenbach D, Finley B [1995]. A study of chromium induced allergic contact dermatitis with 54 volunteers: implications for environmental risk assessment [letter]. Occup Environ Med 52(10):702–704.

Nickens KP, Patierno SR, Ceryak S [2010]. Chromium genotoxicity: a double-edged sword. Chem Biol Interact 188(2):276–288.

NIOSH [1973a]. Criteria for a recommended standard: occupational exposure to chromic acid. Cincinnati, OH: U.S. Department of Health, Education, and Welfare, Public Health Service, Center for Disease Control, National Institute for Occupational Safety and Health, DHEW (NIOSH) Publication No. 73–11021.

NIOSH [1973b]. The industrial environment: its evaluation and control. Washington, DC: U.S. Department of Health, Education, and Welfare, Public Health Service, Center for Disease Control, National Institute for Occupational Safety and Health, DHEW (NIOSH) Publication No. 74–117.

NIOSH [1974]. National occupational hazard survey (NOHS) database, 1972–74. Cincinnati, OH: U.S. Department of Health, Education, and Welfare, Public Health Service, Center for Disease Control, National Institute for Occupational Safety and Health, Division of Surveillance, Hazard Evaluations, and Field Studies, Surveillance Branch, Hazard Section. Unpublished database.

NIOSH [1975a]. Criteria for a recommended standard: occupational exposure to chromium (VI). Cincinnati, OH: U.S. Department of Health, Education, and Welfare, Public Health Service, Center for Disease Control, National Institute for Occupational Safety and Health, DHEW Publication No. (NIOSH) 76–129.

NIOSH [1975b]. Exposure measurement: action level and occupational environmental variability. Cincinnati, OH: U.S. Department of Health, Education, and Welfare, Public Health Service, Center for Disease Control, National Institute for Occupational Safety and Health, DHEW Publication No. (NIOSH) 76–131.

NIOSH [1975c]. Health hazard evaluation determination report: Industrial Platers, Inc., Columbus, Ohio. Cincinnati, OH: U.S. Department of Health, Education, and Welfare, Center for Disease Control, National Institute for Occupational Safety and Health, NIOSH Report No. HHE-74-87-221.

NIOSH [1976]. Current intelligence bulletin 4: chrome pigment. June 24, 1975; October 7, 1975; October 8, 1976. Cincinnati, OH: U.S. Department of Health, Education, and Welfare, Public Health Service, Center for Disease Control, National Institute for Occupational Safety and Health, DHEW (NIOSH) Publication No. 78–127–4.

NIOSH [1977]. Occupational exposure sampling strategy manual. Cincinnati, OH: U.S. Department of Health, Education, and Welfare, Public Health Service, Center for Disease Control, National Institute for Occupational Safety and Health, DHEW (NIOSH) Publication No. 77–173.

NIOSH [1980]. Summarization of recent literature pertaining to an occupational health standard for hexavalent chromium. Rockville, MD: U.S. Department of Health, Education, and Welfare, Public Health Services, Centers for Disease Control, National Institute for Occupational Safety and Health, Contract No. 210-78-0009 for Syracuse Research Corporation, Center for Chemical Hazard Assessment, SRC TR 80–581.

NIOSH [1983a]. National occupational exposure survey (NOES), 1981–83. Cincinnati, OH: U.S. Department of Health and Human Services, Public Health Service, Centers for Disease Control, National Institute for Occupational Safety and Health, Division of Surveillance, Hazard Evaluations, and Field Studies, Surveillance Branch, Hazard Section. Unpublished database.

NIOSH [1983b]. NIOSH comments on the Occupational Safety and Health Administration proposed rule on health standards; methods of compliance: OSHA Docket No. H-160. NIOSH policy statements. Cincinnati, OH: U.S. Department of Health and Human Services, Public Health Service, Centers for Disease Control, National Institute for Occupational Safety and Health.

NIOSH [1985a]. Health hazard evaluation determination report: United Catalysts, Inc. South Plant, Louisville, Kentucky. Cincinnati, OH: Hazard Evaluations and Technical Assistance Branch, U.S. Department of Health and Human Services, Center for Disease Control, National Institute for Occupational Safety and Health. NIOSH Report No. HETA-82-358-1558.

NIOSH [1985b]. Health hazard evaluation determination report: United Catalysts, Inc. West Plant, Louisville, Kentucky. Cincinnati, OH: Hazard Evaluations and Technical Assistance Branch, U.S. Department of Health and Human Services, Center for Disease Control, National Institute for Occupational Safety and Health. NIOSH Report No. HETA 83-075-1559.

NIOSH [1986]. Criteria for a recommended standard: occupational exposure to hot environments. Cincinnati, OH: U.S. Department of Health and Human Services, Public Health Service, Centers for Disease Control, National Institute for Occupational Safety and Health, DHHS (NIOSH) Publication No. 86–113.

NIOSH [1987a]. NIOSH guide to industrial respiratory protection. Cincinnati, OH: U.S. Department of Health and Human Services, Public Health Service, Centers for Disease Control, National Institute for Occupational Safety and Health, DHHS (NIOSH) Publication No. 87–116.

NIOSH [1987b]. NIOSH respirator decision logic. Cincinnati, OH: U.S. Department of Health and Human Services, Public Health Service, Centers for Disease Control and Prevention, National Institute for Occupational Safety and Health, DHHS (NIOSH) Publication No. 87–108.

NIOSH [1988a]. Criteria for a recommended standard: Welding, brazing, and thermal cutting. U.S. Department of Health, Education, and Welfare, Public Health Service, Centers for Disease Control, National Institute for Occupational Safety and Health, Division of Standards Development and Technology Transfer DHHS (NIOSH) Publication No. 88–110.

NIOSH [1988b]. NIOSH testimony on the Occupational Safety and Health Administration's proposed rule on air contaminants, August 1, 1988, OSHA Docket No. H-020. NIOSH policy statements. Cincinnati, OH: U.S. Department of Health and Human Services, Public Health Service, Centers for Disease Control, National Institute for Occupational Safety and Health.

NIOSH [1994a]. Documentation for Immediately Dangerous to Life or Health concentrations. [www.cdc.gov/niosh/idlh]. Date accessed: October 8, 2012.

NIOSH [1994b]. NIOSH manual of analytical methods. 4th ed. Cincinnati, OH: U.S. Department of Health and Human Services, Public Health Service, Centers for Disease Control and Prevention, National Institute for Occupational Safety and Health, DHHS (NIOSH) Publication No. 94–113.

NIOSH [1995a]. Criteria for a recommended standard: occupational exposure to respirable coal mine dust. Cincinnati, OH: U.S. Department of Health and Human Services, Public Health Service Centers for Disease Control and Prevention, National Institute for Occupational Safety and Health, DHHS (NIOSH) Publication No. 95–106.

NIOSH [1995b]. NIOSH Recommended Exposure Limit Policy. Cincinnati, OH: U.S. Department of Health and Human Services, Public Health Service, Centers for Disease Control and Prevention, National Institute for Occupational Safety and Health.

NIOSH [1996a]. Hexavalent chromium in settled dust samples: Method 9101. In: NIOSH manual of analytical methods. 4th ed. Cincinnati, OH: U.S. Department of Health and Human Services, Public Health Service, Centers for Disease Control and Prevention, National Institute for Occupational Safety and Health, DHHS (NIOSH) Publication No. 94–113.

NIOSH [1996b]. NIOSH guide to the selection and use of particulate respirators certified under 42 CFR 84. Cincinnati, OH: U.S. Department of Health and Human Services, Public Health Service, Centers for Disease Control and Prevention, National Institute for Occupational Safety and Health, DHHS (NIOSH) Publication No. 96–101.

NIOSH [1997]. In-depth survey report: control technology assessment for the welding operations at Boilermaker's National Apprenticeship Training School. Cincinnati, OH: U.S. Department of Health and Human Services, Public Health Service, Centers for Disease Control and Prevention, National Institute for Occupational Safety and Health, DHHS (NIOSH) Report No. ECTB 214-13a.

NIOSH [1999]. NIOSH comments on the Occupational Safety and Health Administration proposed rule on employer payment for personal protective equipment: OSHA Docket No. H-042. NIOSH policy statements. Cincinnati, OH: U.S. Department of Health and Human Services, Public Health Service, Centers for Disease Control and Prevention, National Institute for Occupational Safety and Health.

NIOSH [2002]. NIOSH comments on the Occupational Safety and Health Administration request for information on occupational exposure to hexavalent chromium (CrVI): OSHA Docket No. H-0054a. NIOSH policy statements. Cincinnati, OH: U.S. Department of Health and Human Services, Public Health Service, Centers for Disease Control and Prevention, National Institute for Occupational Safety and Health.

NIOSH [2003a]. Hexavalent chromium by field-portable spectrophotometry: Method 7703. In: NIOSH manual of analytical methods. 4th ed. Cincinnati, OH: U.S. Department of Health and Human Services, Public Health Service, Centers for Disease Control and Prevention, National Institute for Occupational Safety and Health, DHHS (NIOSH) Publication No. 94–113.

NIOSH [2003b]. Hexavalent chromium by ion chromatography: Method 7605. In: NIOSH manual of analytical methods. 4th ed. Cincinnati, OH: U.S. Department of Health and Human Services, Public Health Service, Centers for Disease Control and Prevention, National Institute for Occupational Safety and Health, DHHS (NIOSH) Publication No. 94–113.

NIOSH [2003c]. NIOSH pocket guide to chemical hazards. Cincinnati, OH: U.S. Department of Health and Human Services, Centers for Disease Control and Prevention, National Institute for Occupational Safety and Health, NIOSH Publication No. 97–140.

NIOSH [2003d]. Elements on wipes: Method 9102. In: NIOSH manual of analytical methods. 4th ed. Cincinnati, OH: U.S. Department of Health and Human Services, Public Health Service, Centers for Disease Control and Prevention, National Institute for Occupational Safety and Health, DHHS (NIOSH) Publication No. 94–113.

NIOSH [2004]. NIOSH respirator selection logic. Cincinnati, OH: U.S. Department of Health and Human Services, Public Health Service, Centers for Disease Control and Prevention, National Institute for Occupational Safety and Health, DHHS (NIOSH) Publication No. 2005–100 [http://www.cdc.gov/niosh/docs/2005-100].

NIOSH [2005a]. NIOSH testimony on the Occupational Safety and Health Administration's proposed rule on occupational exposure to hexavalent chromium, January 5, 2005, OSHA Docket No. H-054A. NIOSH policy statements. Cincinnati, OH: U.S. Department of Health and Human Services, Public Health Service, Centers for Disease Control and Prevention, National Institute for Occupational Safety and Health.

NIOSH [2005b]. NIOSH posthearing comments on the Occupational Safety and Health Administration's proposed rule on occupational exposure to hexavalent chromium, March 21, 2005, OSHA Docket No. H-054A. NIOSH policy statements. Cincinnati, OH: U.S. Department of Health and Human Services, Public Health Service, Centers for Disease Control and Prevention, National Institute for Occupational Safety and Health.

NIOSH [2006]. NIOSH criteria for a recommended standard: occupational exposure to refractory ceramic fibers. Cincinnati, OH: U.S. Department of Health and Human Services, Centers for Disease Control and Prevention, National Institute for Occupational Safety and Health, DHHS (NIOSH) Publication No. 2006–123.

NIOSH [2007]. NIOSH hexavalent chromium topic page. [http://www.cdc.gov/niosh/topics/hexchrom/]. Date accessed: November 15, 2007.

NIOSH [2008a]. NIOSH protective clothing and ensembles topic page [http://www.cdc.gov/niosh/topics/protclothing/]. Date accessed: September 1, 2008.

NIOSH [2008b]. NIOSH respirators topic page [http://www.cdc.gov/niosh/topics/respirators/]. Date accessed: September 1, 2008.

NIOSH [2011]. NIOSH spirometry topic page. [http://www.cdc.gov/niosh/topics/spirometry/]. Date accessed: November 1, 2011.

NTP [1996a]. Final report on the reproductive toxicity of potassium dichromate (hexavalent) (CAS No. 7778-50-9) administered in diet to SD rats. National Institute of Environmental Health Sciences, National Toxicology Program. PB97125355.

NTP [1996b]. Final report on the reproductive toxicity of potassium dichromate (hexavalent) (CAS No. 7778-50-9) administered in diet to BALB/c mice. National Institute of Environmental Health Sciences, National Toxicology Program. DHHS (NIH) Publication No. PB97125363.

NTP [1997]. Final report on the reproductive toxicity of potassium dichromate (CAS No. 7778-50-9) administered in diet to BALB/c mice. National Institute of Environmental Health Sciences, National Toxicology Program. DHHS (NIH) Publication No. PB97144919.

NTP [2011]. Report on Carcinogens, Twelfth Edition. Research Triangle Park, NC: U.S. Department of Health and Human Services, Public Health Service, National Institutes of Health, National Toxicology Program.

OEHHA [2009]. Evidence on the developmental and reproductive toxicity of chromium (hexavalent compounds). Reproductive and Cancer Hazard Assessment Branch, Office of Environmental Health Hazard Assessment, California Environmental Protection Agency [http://www.oehha.ca.gov/prop65/hazard_ident/pdf_zip/chrome0908.pdf].

Okubo T, Tsuchiya K [1977]. An epidemiological study on lung cancer among chromium plating workers. Keio J Med 26:171–177.

Okubo T, Tsuchiya K [1979]. Epidemiological study of chromium platers in Japan. Biol Trace Element Res 1:35–44.

Okubo T, Tsuchiya K [1987]. Mortality determined in a by cohort study of chromium-plating workers [abstract]. Scand J Work Environ Health 13:179.

OSHA [1998]. Hexavalent chromium in workplace atmospheres: Method ID-215. In: OSHA Analytical Methods Manual. Salt Lake City, UT: U.S. Department of Labor, Occupational Safety and Health Administration.

OSHA [1999a]. Chemical protective clothing. In: OSHA Technical Manual, Section VIII, Chapter 1. Office of Science and Technology Assessment: U.S. Department of Labor, Occupational Safety and Health Administration.

OSHA [1999b]. Metals sampling. In: OSHA Technical Manual, Section II, Chapter 1. Office of Science and Technology Assessment: U.S. Department of Labor, Occupational Safety and Health Administration.

OSHA [2001]. Hexavalent chromium. Wipe sampling method: Method W-4001. In: OSHA analytical methods manual. Salt Lake City, UT: U.S. Department of Labor, Occupational Safety and Health Administration.

OSHA [2006]. Hexavalent chromium: Method ID-215 (Version 2). In: OSHA Analytical Methods Manual. Salt Lake City, UT: U.S. Department of Labor, Occupational Safety and Health Administration.

OSHA [2007]. OSHA safety and health topic: OSHA standards: hexavalent chromium. [http://www.osha.gov/SLTC/hexavalentchromium/index.html]. Date accessed: November 15, 2007.

OSHA [2008]. Preventing skin problems from working with Portland cement. Washington, DC: U.S. Department of Labor, Occupational Safety and Health Administration, OSHA Publication No. 3351-07.

OSHA [2012]. SIC Manual (1987) Web page. [www.osha.gov/pls/imis/sic_manual.html]. Date accessed: October 8, 2012.

Paddle GM [1997]. Metaanalysis as an epidemiological tool and its application to studies of chromium. Reg Toxicol Pharmacol 26(1 Pt 2):S42–S50.

Park RM, Stayner LS [2006]. A search for thresholds and other non-linearities in the relationship between hexavalent chromium and lung cancer. Risk Anal 26(1):79–88.

Park RM, Bena JF, Stayner LT, Smith RJ, Gibb HJ, Lees PS [2004]. Hexavalent chromium and lung cancer in the chromate industry: a quantitative risk assessment. Risk Anal 24(5):1099–1108.

Park RM, Maizlish NA, Punnett L, Moure-Eraso R, Silverstein MA [1991]. A comparison of PMRs and SMRs as estimators of occupational mortality. Epidemiology 2(1):49–59.

Park R, Rice F, Stayner L, Smith R, Gilbert S, Checkoway H [2002]. Exposure to crystalline silica, silicosis, and lung disease other than cancer in diatomaceous earth industry workers: a quantitative risk assessment. Occup Environ Med 59:36–43.

Park HS, Yu HJ, Jung KS [1994]. Occupational asthma caused by chromium. Clin Exp Allergy 24(7):676–681.

Pastides H, Austin R, Lemeshow S, Klar J, Mundt KA [1994a]. A retrospective-cohort study of occupational exposure to hexavalent chromium. Am J Ind Med 25(5):663–675.

Pastides H, Austin R, Mundt KA, Ramsey F, Feger N [1994b]. Transforming industrial hygiene data for use in epidemiologic studies: a case study of hexavalent chromium. J Occup Med Toxicol 3(1):57–71.

Paustenbach DJ, Sheehan PJ, Paull JM, Wisser LM, Finley BL [1992]. Review of the allergic contact dermatitis hazard posed by chromium-contaminated soil: identifying a "safe" concentration. J Toxicol Environ Health 37(1):177–207.

Pokrovskaya LV, Shabynina NK [1973]. Carcinogenous hazards in the production of chromium ferroalloys. Gig Tr Prof Zabol 17(10):23–26.

Polak L [1983]. Immunology of chromium. In: Burrows D, ed. Chromium: metabolism and toxicity. Boca Raton, FL: CRC Press, Inc., pp. 51–136.

Proctor DM, Fredrick MM, Scott PK, Paustenbach DJ, Finley BL [1998]. The prevalence of chromium allergy in the United States and its implications for setting soil cleanup: a cost-effectiveness case study. Regul Toxicol Pharmacol 28(1):27–37.

Proctor DM, Otani JM, Finley BL, Paustenbach DJ, Bland JA, Speizer N, Sargent EV [2002]. Is hexavalent chromium carcinogenic via ingestion? A weight-of-evidence review. J Toxicol Environ Health A 65(10):701–746.

Proctor DM, Panko JP, Leibig EW, Scott PK, Mundt KA, Buczynski MA, Barnhart RJ, Harris MA, Morgan RJ, Paustenbach DJ [2003]. Workplace airborne hexavalent chromium concentrations for the Painesville, Ohio, chromate production plant (1943–1971). Appl Occup Environ Hyg 18(6):430–449.

Proctor DM, Panko JP, Liebig EW, Paustenbach DJ [2004]. Estimating historical occupational exposure to airborne hexavalent chromium in a chromate production plant: 1940–1972. J Occup Environ Hyg 1(11):752–767.

Qian Y, Jiang B, Flynn DC, Leonard SS, Wang S, Zhang Z, Ye J, Chen F, Wang E, Shi X [2001]. Cr(VI) increases tyrosine phosphorylation through reactive oxygen species-mediated reacrions. Mol Cell Biochem 222(1–2):199–204.

Rafnsson V, Jóhannesdóttir SG [1986]. Mortality among masons in Iceland. Br J Ind Med 43:522–525.

Rice FL, Park R, Stayner L, Smith R, Gilbert S, Checkoway H [2001]. Crystalline silica exposure and lung cancer mortality in diatomaceous earth industry workers: a quantitative risk assessment. Occup Environ Med 58(1):38–45.

Rosenman KD, Stanbury M [1996]. Risk of lung cancer among former chromium smelter workers. Am J Ind Med 29(5):491–500.

Roto P, Sainio H, Renuala T, Laippala P [1996]. Addition of ferrous sulfate to cement and risk of chromium dermatitis among construction workers. Contact Dermatitis 34:43–50.

Royle H [1975a]. Toxicity of chromic acid in the chromium plating industry (1). Environ Res 10(1):39–53.

Royle H [1975b]. Toxicity of chromic acid in the chromium plating industry (2). Environ Res 10(1):141–163.

Rudolf E, Cervinka M, Cerman J, Schroterova L [2005]. Hexavalent chromium disrupts the actin cytoskeleton and induces mitochondria-dependent apoptosis in human dermal fibroblasts. Toxicol In Vitro 19(6):713–723.

Satoh K, Eng B, Fukuda Y, Kazuyoshi T, Eng M, Katsuno N [1981]. Epidemiological study of workers engaged in the manufacture of chromium compounds. J Occup Med 23(12):835–838.

Satoh N, Fukuda S, Takizawa M, Furuta Y, Kashiwamura M, Inuyama Y [1994]. Chromium-induced carcinoma in the nasal region. A report of four cases. Rhinology 32(1):47–50.

Scheepers PT, Heussen GA, Peer PG, Verbist K, Anzion R, Willems J [2008]. Characterisation of exposure to total and hexavalent chromium of welders using biological monitoring. Toxicol Lett 178(3):185–190.

Schulte PA [1995]. Opportunities for the development and use of biomarkers. Toxicol Lett 77:25–29.

Shaw Environmental [2006]. Industry profile, exposure profile, technological feasibility evaluation, and environmental impact for industries affected by a proposed OSHA standard for hexavalent chromium. Cincinnati, OH: Shaw Environmental, Inc. Contract No. J-9-F-9-0030, Subcontract No. 0178.03.062/1, PN 118851-01 for OSHA, U.S. Department of Labor.

Sheffet A, Thind I, Miller AM, Louria DB [1982]. Cancer mortality in a pigment plant utilizing lead and zinc chromates. Arch Environ Health 37(1):44–52.

Shi X, Dalal NS [1990a]. Evidence for a Fenton-type mechanism for the generation of ·OH radicals in the reduction of Cr(VI) in cellular media. Arch Biochem Biophys 281(1):90–95.

Shi X, Dalal NS [1990b]. NADPH-dependent flavoenzymes catalyze one electron reduction of metal ions and molecular oxygen and generate hydroxyl radicals. FEBS Lett 276(1–2):189–191.

Shi X, Dalal NS [1990c]. On the hydroxyl radical formation in the reaction between hydrogen peroxide and biologically generated chromium (V) species. Arch Biochem Biophys 277(2):342–350.

Shi X, Mao Y, Knapton AD, Ding M, Rojanasakul Y, Gannett PM, Dalal N, Liu K [1994]. Reaction of Cr(VI) with ascorbate and hydrogen peroxide generates hydroxyl radicals and causes DNA damage: role of Cr(VI)-mediated Fenton-like reaction. Carcinogenesis 15(11):2475–2478.

Shmitova LA [1978]. The course of pregnancy in women engaged in the production of chromium and its compounds. Vliy Prof Fakt Spet Funk Zhensk Organ, Sverdl: 19:108–111.

Shmitova LA [1980]. Content of hexavalent chromium in the biological substrates of pregnant women and women in the immediate post-natal period engaged in the manufacture of chromium compounds. Gig Tr Prof Zabol 2(2):33–35.

Silverstein M, Mirer F, Kotelchuck D, Silverstein B, Bennett M [1981]. Mortality among workers in a die-casting and electroplating plant. Scand J Work Environ Health 7(Suppl 4):156–165.

Simonato L, Fletcher AC, Andersen A, Anderson K, Becker N, Chang-Claude J, Ferro G, Gerin M, Gray CN,

Hansen KS, Kalliomäki P-L, Kurppa K, Langård S, Merló F, Moulin JJ, Newhouse ML, Peto J, Pukkala E, Sjogren B, Wild P, Winkelmann R, Saracci R [1991]. A historical prospective study of European stainless steel, mild steel and shipyard welders. Br J Ind Med 48(3):145–154.

Singh J, Carlisle DL, Pritchard DE, Patierno SR [1998]. Chromium-induced genotoxicity and apoptosis: relationship to chromiom carcinogenesis (review). Oncol Rep 5(6):1307–1318.

Sjögren B, Gustavsson A, Hedström L [1987]. Mortality in two cohorts of welders exposed to high- and low-levels of hexavalent chromium. Scand J Work Environ Health 13(3):247–251.

Sjögren B, Hansen KS, Kjuus H, Persson P-G [1994]. Exposure to stainless steel welding fumes and lung cancer: a meta-analysis. Occup Environ Med 51(5):335–336.

Slyusar TA, Yakovlev NA [1981]. Clinical features of cerebral arachnoiditis in chromium industry workers. Zdravookhranenie Kazakhstana 10:32–34.

Sorahan T, Burges DCL, Waterhouse JAH [1987]. A mortality study of nickel/chromium platers. Br J Ind Med 44(4):250–258.

Sorahan T, Harrington JM [2000]. Lung cancer in Yorkshire chrome platers, 1972-97. Occup Environ Med 57(6):385–389.

Soule RD [1978]. Industrial Hygiene engineering controls. Chapter 18. In: Clayton GD, Clayton FE, eds. Pattys industrial hygiene and toxicology. 3rd rev ed. Vol. I. General Principles. New York: Wiley Interscience, John Wiley & Sons.

Stayner LT, Dankovic DA, Smith RJ, Gilbert SJ, Bailer AJ [2000]. Human cancer risk and exposure to 1,3-butadiene—a tale of mice and men. Scand J Work Environ Health 26(4):322–330.

Steenland K, Loomis D, Shy C, Simonsen N [1996]. Review of occupational lung carcinogens. Am J Ind Med 29(5):474–490.

Steinhoff D, Gad SC, Hatfield GK, Mohr U [1986]. Carcinogenicity studies with sodium dichromate in rats. Exp Pathol 30(3):129–141.

Sterekhova NP, Zeleneva NI, Solomina SN, Tiushniakova NV, Miasnikova AG, Fokina GP, Yarina AL [1978]. Gastric pathology in workers engaged in the production of chromium salts. Gig Tr Prof Zabol 3:19–23.

Stern AH, Bagdon RE, Hazen RE, Marzulli FN [1993]. Risk assessment of the allergic dermatitis potential of environmental exposure to hexavalent chromium. J Toxicol Environ Health 40(4):613–641.

Stern FB [2003]. Mortality among chrome leather tannery workers: an update. Am J Ind Med 44(2):197–206.

Stern FB, Beaumont JJ, Halperin WE, Murthy LI, Hills BW, Fajen JM [1987]. Mortality of chrome leather tannery workers and chemical exposures in tanneries. Scand J Work Environ Health 13(2):108–117.

Sterns DM, Kennedy LJ, Courtney KD, Giangrande PH, Phieffer LS, Wetterhahn KE [1995]. Reduction of chromium(VI) by ascorbate leads to chromium-DNA binding and DNA strands breaks in vitro. Biochemistry 34(3):910–919.

Stoner RS, Tong TG, Dart R, Sullivan JB, Saito G, Armstrong E [1988]. Acute chromium intoxication with renal failure after 1% body surface area burns from chromic acid [abstract]. Vet Human Toxicol 30(4):361–362.

Straif K, Benbrahim-Tallaa L, Baan R, Grosse Y, Secretan B, El Ghissassi F, Bouvard V, Guha N, Freeman C, Galichet L, Cogliano V [2009]. A review of human carcinogens—part C: metals, arsenic, dusts, and fibres. Lancet Oncol 10(5):453–454.

Sugiyama M, Wang X, Costa M [1986]. Comparison of DNA lesions and cytotoxicity induced by calcium chromate in human, mouse and hamster cell lines. Cancer Res 46(9):4547–4551.

Svensson BG, Englander V, Åkesson B, Attewell R, Skerfving S, Ericson Å, Möller T [1989]. Deaths and tumors among workers grinding stainless steel. Am J Ind Med 15(1):51–59.

Takahashi K, Okubo T [1990]. A prospective cohort study of chromium plating workers in Japan. Arch Environ Health 45(2):107–111.

Taylor FH [1966]. The relationship of mortality and duration of employment as reflected by a cohort of chromate workers. Am J Public Health Nations Health 56(2):218–229.

Tsapakos MJ, Wetterhahn KE [1983]. The interaction of chromium with nucleic acids. Chem Biol Interact 46(2):265–277.

Tsapakos MJ, HamptonTH, Wetterhahn KE [1983]. Chromium (IV)-induced DNA lesions and chromium distribution in rat kidney, liver and lung. Cancer Res 43(12 Pt 1):5662–5667.

USGS [2009]. Minerals Yearbook: Chromium [http://minerals.usgs.gov/minerals/pubs/commodity/chromium]. Date accessed: August 31, 2011.

USGS [2012]. Mineral Commodity Summary: Chromium [http://minerals.usgs.gov/minerals/pubs/commodity/chromium]. Date accessed: September 19, 2012.

Van Lierde V, Chery CC, Roche N, Monstrey S, Moens L, Vanhaecke F [2006]. In vitro permeation of chromium species through porcine and human skin as determined by capillary electrophoresis-inductively coupled plasma-sector field mass spectrometry. Anal Bioannal Chem 384(2):378–384.

Van Wijngaarden E, Mundt KA, Luippold RS [2004]. Evaluation of the exposure-response relationship of lung cancer mortality and occupational exposure to hexavalent chromium based on published epidemiological data. Nonlinearity Biol Toxicol Med 2(1):27–34.

Wang J, Ashley K, Marlow D, England EC, Carlton G [1999]. Field method for the determination of hexavalent chromium by ultrasonication and strong anion-exchange solid-phase extraction. Anal Chem 71(5):1027–1032.

Wang S, Shi X [2001]. Mechanism of Cr(VI)-induced p53 activation: the role of phosphorylation, mdm2 and ERK. Carcinogenesis 22:757–762.

Wang S, Leonard SS, Ye J, Ding M, Shi X [2000]. The role of hydroxyl radical as a messenger in Cr(VI)-induced p53 activation. Am J Physiol 279:C868–C875.

Wang S, Leonard S, Ye J, Ding M, Shi X [2000]. The role of hydroxyl radicals as messenger on Cr (VI)-induced p53 activation. Am J Physiol Cell Physiol 279(3):C868–C875.

Watanabe S, Fukuchi Y [1984]. Cancer mortality of chromate-producing workers [abstract]. In: Eustace IE, ed. XXI International Congress on Occupational Health, September 9–14, 1984, Dublin, Ireland: Permanent Commission and International Association on Occupational Health, London, p. 442.

Waterhouse JAH [1975]. Cancer among chromium platers. Br J Cancer 32(2):262.

Weber H [1983]. Long-term study of the distribution of soluble chromate-51 in the rat after a single intratracheal administration. J Toxicol Environ Health 11(4–6):749–764.

WHO [1988]. Chromium. Environmental Health Criteria 61. Geneva, Switzerland: World Health Organization, International Programme on Chemical Safety (IPCS) [http://www.inchem.org/documents/ehc/ehc/ehc61.htm]. Date accessed: June 16, 2003.

Wiegand HJ, Ottenwalder H, Bolt HM [1985]. Fast uptake kinetics in vitro of Cr(VI) by red blood cells of man and rat. Arch Toxicol 57:31–34.

Wiegand HJ, Ottenwalder H, Bolt HM [1988]. Recent advances in biological monitoring of hexavalent chromium compounds. Sci Total Environ 71:309–315.

Wise JP, Leonard JC, Patierno SR [1992]. Clastogenicity of lead chromate particles in hamster and human cells. Mut Res 278(1):69–79.

Wise JP, Orenstein JM, Patierno SR [1993]. Inhibition of lead chromate clastogenesis by ascorbate: relationship to particle dissolution and uptake. Carcinogenesis 14(3):429–434.

Wise JP Sr., Stearns DM, Wetterhahn KE, Patierno SR [1994]. Cell-enhanced dissolution of carcinogenic lead chromate particles: the role of individual dissolution products in clastogenesis. Carcinogenesis 15(10):2249–2254.

Wise JP Sr., Wise SS, Little JE [2002]. The cytotoxicity and genotoxicity of particulate and soluble hexavalent chromium in human lung cells. Mutat Res 517(1–2):221–229.

Wise SS, Holmes AL, Wise JP Sr. [2006a]. Particulate and soluble hexavalent chromium are cytotoxic and genotoxic to human lung epithelial cells. Mut Res 610(1–2):2–7.

Wise SS, Holmes AL, Xie H, Thompson WD, Wise JP Sr. [2006b]. Chronic exposure to particulate chromate induces spindle assembly checkpoint bypass in human lung cells. Chem Res Toxicol 19(11):1492–1498.

Wise SS, Schuler JH, Katsifis SP, Wise JP Sr. [2003]. Barium chromate is cytotoxic and genotoxic to human lung cells. Environ Mol Mutagen 42(4):274–278.

Xie H, Holmes AL, Wise SS, Gordon N, Wise JP Sr. [2004]. Lead chromate-induced chromosome damage requires extracellular dissolution to liberate chromium ions but does not require particle internalization or intracellular dissolution. Chem Res Toxicol 17(10):1362–1367.

Xie H, Holmes AL, Wise SS, Huang S, Peng C, Wise JP Sr. [2007]. Neoplastic transformation of human bronchial cells by lead chromate particles. Am J Resp Cell Mol Bio *37*(5):544–552.

Xie H, Wise SS, and Wise JP Sr. [2008]. Deficient repair of particulate chromate-induced DNA double strand breaks leads to neoplastic transformation. Mut Res *649*:230–238.

Xie H, Holmes AL, Young JL, Qin Q, Joyce K, Pelsue SC, Peng C, Wise SS, Jeevarajan A, Wallace WT, Hammond D, Wise JP Sr. [2009]. Zinc chromate induces chromosome instability and DNA double strand breaks in human lung cells. Toxicol App Pharmacol *234*(3):293–299.

Ye J, Shi X [2001]. Gene expression profile in response to chromium induced cell stress in A549 cells. Mol Cell Biochem *222*(1–2):189–197.

Ye J, Wang S, Leonard SS, Sun Y, Butterworth L, Antonini J, Ding M, Rojanasakul Y, Vallyathan V, Castranova V, Shi X [1999]. Role of reactive oxygen species and p53 in chromium(VI)-induced apoptosis. J Biol Chem *274*(49):34974–34980.

Zhang Z, Leonard SS, Wang S, Vallyathan V, Castranova V, Shi X [2001]. Cr(VI) induces cell growth arrest through hydrogen peroxide-mediate reactions. Mol Cell Biochem *222*(1–2):77–83.

Appendix A

Hexavalent Chromium and Lung Cancer in the Chromate Industry: A Quantitative Risk Assessment

Hexavalent Chromium and Lung Cancer in the Chromate Industry: A Quantitative Risk Assessment*

Robert M. Park[†]
James F. Bena[†]
Leslie T. Stayner[†]
Randall J. Smith[†]
Herman J. Gibb[‡]
Peter S.J. Lees[§]

---

*This manuscript was submitted for publication to the peer-reviewed journal Risk Analysis. For the final publication see: Park RM, Bena JF, Stayner LT, Smith RJ, Gibb HJ, Lees PS [2004]. Hexavalent chromium and lung cancer in the chromate industry: a quantitative risk assessment. Risk Anal 24(5):1099–1108.

[†]U.S. Department of Health and Human Services, Public Health Service, Centers for Disease Control and Prevention, National Institute for Occupational Safety and Health, 4676 Columbia Parkway, MS C-15, Cincinnati, OH 45226-1998, USA; Phone: 513-533-8572; E-mail: rhp9@cdc.gov

[‡]U.S. Environmental Protection Agency, Washington, DC

[§]The Johns Hopkins University Bloomberg School of Public Health, Baltimore, MD

ABSTRACT

Objectives: The purpose of this investigation was to estimate excess lifetime risk of lung cancer death resulting from occupational exposure to hexavalent chromium-containing dusts and mists.

Methods: The mortality experience in a previously studied cohort of 2357 chromate chemical production workers with 122 lung cancer deaths was analyzed with Poisson regression methods. Extensive records of air samples evaluated for water-soluble total hexavalent chromium were available for the entire employment history of this cohort. Six different models of exposure-response for hexavalent chromium were evaluated by comparing deviances and inspection of cubic splines. Smoking (pack-years) imputed from cigarette use at hire was included in the model. Lifetime risks of lung cancer death from exposure to hexavalent chromium (assuming up to 45 years of exposure) were estimated using an actuarial calculation that accounts for competing causes of death.

Results: A linear relative rate model gave a good and readily interpretable fit to the data. The estimated rate ratio for 1 mg/m^3-yr of cumulative exposure to hexavalent chromium (as CrO_3), with a lag of 5 years, was RR = 2.44 (95% CI=1.54–3.83). The excess lifetime risk of lung cancer death from exposure to hexavalent chromium at the current OSHA Permissible Exposure Limit (0.10 mg/m^3) was estimated to be 255 per 1000 (95% CI: 109–416). This estimate is comparable to previous estimates by U.S. EPA, California EPA and OSHA using different occupational data.

Conclusions: Our analysis predicts that current occupational standards for hexavalent chromium permit a lifetime excess risk of dying of lung cancer that exceeds 1 in 10, which is consistent with previous risk assessments.

Keywords: Excess lifetime risk, hexavalent chromium exposure-response, race interaction

INTRODUCTION

History of Hexavalent Chromium Hazard

Chromium is commercially important in metallurgy, electroplating, and in diverse chemical applications such as pigments, biocides and strong oxidizing agents. Adverse health effects have long been known and include skin ulceration, perforated nasal septum, nasal bleeding, and conjunctivitis. Reports of bronchogenic carcinoma appeared prior to World War II in Germany and were subsequently confirmed in multiple studies.[1] The International Agency for Research on Cancer (IARC) declared in 1980 that chromium and certain of its compounds are carcinogenic and, in 1987, concluded that hexavalent chromium is a human carcinogen but that trivalent chromium was not yet classifiable. Recent studies updating chromium worker cohorts in Ohio[2,3] and Maryland[1] demonstrated an excess lung cancer risk from exposure to hexavalent chromium.

Regulation of Chromium Exposures

The current Permissible Exposure Limit (PEL) of the U.S. Occupational Safety and Health Administration (OSHA) is 0.1 mg/m^3 for soluble hexavalent chromium (as CrO_3) as an 8 hour time-weighted average.[4] The American Conference of Governmental Industrial Hygienists has a similar recommendation.[5] The U.S. National Institute for Occupational Safety and Health recommends a limit of 0.001 mg/m^3 (as Cr).[6] Due to continuing concerns over lung cancer risks from hexavalent chromium, OSHA is currently reviewing the distribution of chromium exposures in the U.S. workforce and the available estimates of excess risk.

Present Objective

The goal of this investigation was to evaluate various models of exposure-response for lung cancer mortality and exposure to hexavalent chromium compounds and then conduct a risk assessment for lung cancer based on these models. The cohort of chromate workers analyzed by Hayes et al.[7] and later updated and modified by Gibb et al.[1] was used for the analysis. In addition to a detailed retrospective exposure assessment for chromium, this cohort had smoking information and is believed to be largely free of other potentially confounding exposures from this plant. Using log-transformed cumulative exposure estimates within a proportional hazards regression model, Gibb et al. observed the rate of lung cancer mortality in these chromate workers to increase by a factor of 1.38 for each 10-fold increase in cumulative exposure to hexavalent chromium (p=0.0001).[1]

METHODS

The description of the cohort can be found in Gibb et al. and comprised 2372 men hired between August 1, 1950 and December 31, 1974 at a plant in Baltimore, MD.[1,7] Fifteen workers lost to followup were excluded, leaving 2357 subjects for analysis. Followup began with date of hire and continued until December 31, 1992 or the date of death, whichever occurred first. The cohort consisted of 1205 men known to be white (51%), 848 known to be nonwhite (36%)—believed to be mostly African Americans—and 304 with unknown race (13%). The mean duration of employment was 3.1 yr., but the median was 0.39 yr. Some smoking information was available at hire for 91% of the study population, including smoking level in packs per day for 70%. For those of unknown smoking status (9%), average levels were assigned; for known cigarette smokers with unknown cigarette usage (21%), the average level among known smokers was assigned. Cumulative smoking exposure, as packs/day–years, was calculated assuming workers smoked from age 18 until the end of followup, and using a 5 year lag (same as for chromium exposure).

Chromium Exposure History

This chromate manufacturing facility began operation in 1845 and continued until 1985. Because of facility and process changes and the limited availability of detailed early air sampling data, the study population was restricted to those who worked in the "new" plant and were hired between August 1, 1950 and December 31, 1974. A detailed retrospective exposure assessment was conducted for this population[1] using contemporaneous exposure measurements. During 1950–61, short-term personal samples were collected using high volume pumps. From 1961 until 1985 approximately 70,000 systematic area air samples were collected at 154 fixed sites (27 sites after 1977). Based on these air samples and recorded observations of the fraction of time spent in these exposure zones by each job title, the employer calculated exposures by job title. After 1977, full-shift personal samples were collected as well. Hexavalent chromium concentrations (as CrO₃) were based on laboratory determinations of water-soluble chromate performed by the employer; results of the area sampling/time-in-zone system of calculating exposures were adjusted to the personal sample results. Exposure histories were then calculated for each worker based on the jobs held (defined by dates) and corresponding hexavalent chromium exposure estimate for that job title and time period. Total cumulative exposures to hexavalent chromium averaged 0.134 mg/m³-yr, with a maximum value of 5.3 mg/m³-yr.[1]

Mortality Analyses

A classification table for Poisson Regression analysis was calculated using a FORTRAN program developed previously[8] which classified followup in 16 age (<20, 20–24, 25–29, 30–34,... , 85+), 9 calendar (1950–54, 1955–59,...1990–94), and three race categories (0=nonwhite, 1=white, 2= unknown), together with 50 levels of time-dependent cumulative hexavalent chromium exposure and 10 levels of time-dependent cumulative smoking exposure and employment duration. The 50 intervals of cumulative exposure were defined to be narrower at the low exposure end compared to the high end because observation time was concentrated at the low end. Although classified in discrete levels, cumulative exposures for chromium and smoking were entered into regression models as a continuous variable defined by the person-year weighted mean values of cumulative exposure in each of the classification strata. Cumulative exposure is a standard metric used in modeling carcinogenic risk in human populations. The unit of followup in this procedure was 30 days, i.e., every 30-day interval of a worker's observed person-time (and lung cancer death outcome) was classified as described above. To address latency, different lag periods were used and, for some analyses, cumulative exposures were calculated in three time periods: 5–9.99, 10–19.99, 20 or more years prior to observation.

Relative rate models of the following forms (1a–1f) were evaluated for the effect of cumulative chromium exposure on lung cancer mortality:

Log-linear models

Log-linear: $\text{rate} = \exp(â_0 + âX)$ or $\ln[\text{rate}] = â_0 + âX$ \hfill (1a)

Log-square root: $\text{rate} = \exp(â_0 + âX^{0.5})$ \hfill (1b)

Log-quadratic: $\text{rate} = \exp(â_0 + â_1 X + â_2 X^2)$ \hfill (1c)

Power: $\text{rate} = \exp(â_0 + â \ln[X+1]) = \exp(â_0) \times (X+1)^â$ \hfill (1d)

Additive relative rate models

Linear relative rate: $\text{rate} = \exp(â_0) \times (1 + âX)$ \hfill (1e)

"Shape": $\text{rate} = \exp(â_0) \times (1 + â_1 X^{â_2})$ \hfill (1f)

where, $â_0$ (intercept), $â$, $â_1$, and $â_2$ are parameters to be estimated.

External standardization on age, race and calendar time was accomplished using U.S. rates for lung cancer mortality during 1950–1994 [9] as a multiplier of person-years. This approach makes use of well-known population rates and yields models of standardized rate ratios in which the intercept is an estimate of the (log) standardized mortality ratio (SMR) for workers without chromium or smoking exposures. The method permits departures from the reference rates by including explicit terms (e.g., age or race) in the model, and it enables internal comparisons on exposure. Those with unknown race (n=304) were assigned a composite expected rate as a weighted average of the external race-specific rates based on the distribution of race in their year of hire.

Poisson regression models were fit using Epicure software.[10] Descriptive categorical analyses for chromium exposure-response were conducted using five levels of cumulative exposure defined to produce equal numbers of lung cancer cases in the upper levels for both races combined. The lowest category encompassed the most observation time and observed deaths because of the large number of short duration employees in this study population (median duration: 4.7 mos.). Files were also constructed for Cox proportional hazards models with continuous cumulative exposures and risk sets based on the attained age of the lung cancer case at death. Results using Cox proportional hazard analyses were similar to the Poisson analyses and are not shown.

The models with the largest decrease in deviance (i.e., decrease in −2log (likelihood) with addition of exposure terms) were considered to be the "best" fitting. The adequacy of the parametric forms for exposure and for smoking duration in these models was also investigated by fitting cubic splines[11]. The spline models were fit as generalized additive models with three degrees of freedom for the effect of exposure using S-Plus, version 4.5. [12]

Unlagged and lagged cumulative exposures were considered. Models with exposures lagged by 5 yrs or 10 yrs provided statistically equivalent fits to the data (based on minimizing the deviance) which were better than that obtained with the unlagged model; a 5 yr lag was chosen to conform to previous analyses of this cohort.[1] The cumulative smoking variable was included in the log-linear or linear terms of different models. However, in estimating the chromium effect for calculating excess lifetime risk, the smoking variable was placed in the loglinear term in order to produce a chromium risk estimate relative to an unexposed population without regard to smoking status. Finally, to better model the smoking exposure-response, a piece-wise linear spline was used with a knot chosen at 30 packs/day–yrs). This choice was motivated by a cubic spline analysis of the smoking effect exhibiting a plateau above 30 pack-years. The final model chosen for estimating excess lifetime lung cancer risk had the following form, consisting of the product of "loglinear" and "linear" terms:

$$\text{Rate} = [\exp(a_0 + a_1 \text{Ind}(w) + a_2 \text{Ind}(\text{unk}) + a_3(\text{Age}-50) + a_4 \text{Smk1} + a_5 \text{Smk2})] \times [1 + b_1 \text{CumCr6}]$$

where: Ind() are indicators of race (white or unknown), Smk1 and Smk2 are variables specified for the piece-wise linear terms on cumulative smoking, and CumCr6 is cumulative hexavalent chromium exposure.

Estimation of Working Lifetime Risks

Excess lifetime risk of death from lung cancer was estimated for a range of chromium air concentrations using an actuarial method that accounts for competing risks and was originally developed for a risk analysis of radon.[13] Excess lifetime risk was estimated by first applying cause-specific rates from an exposure-response model to obtain lifetime risk, and then subtracting the same expression with exposures set to zero:

Excess lifetime risk =

$\Sigma_i \{ [R_i(X)/R_{+i}(X)] \times S(X,i) \times [q_i(X)] \} - \Sigma_i \{ [R_i(0)/R_{+i}(0)] \times S(0,i) \times [q_i(0)] \}$

where $R_{+i}(X)$ = all cause age-specific mortality rate for exposed population; q_i = Pr(death in year i given alive at the start of year i); and $S(X,i) = (1-q_1) \times (1-q_2) \times \ldots \times (1-q_{i-1})$, probability of survival to year i

For specified hexavalent chromium concentrations, excess lifetime risks were estimated making the assumption that workers were occupationally exposed to constant concentrations between the ages of 20 and 65, or 45 years (less if dying before age 65). Annual excess risks were accumulated up to age 85; risk among those surviving past age 85 was not calculated because of small numbers and unstable rates. Rate ratios for lung cancer mortality corresponding to work at various chromium concentrations were derived from the final linear relative rate Poisson regression model. Age-specific all-cause death rates came from a life table for the U.S. population.[14]

RESULTS

Lung Cancer Mortality

Lung cancer was the underlying cause for 122 deaths in the chromate cohort. Fitting a Poisson regression model with indicator terms for race produced similar lung cancer SMRs for white (SMR=1.85, 95%CI=1.45–2.31) and nonwhite (SMR=1.87, 95%CI=1.39–2.46) workers that were close to those reported by Gibb et al. (1.86, 1.88, respectively, using different reference rates).[1]

Results from fitting models with five categorical chromium exposure levels and unadjusted for smoking reveal a clear upward trend for the lung cancer SMR with cumulative chromium exposure but the trends differed by race (Table 1). The same patterns were observed using internally standardized rate ratios (SRRs, Table 1). The nonwhite workers exhibited a strong overall increasing trend of lung cancer risk except for a deficit in the 2nd exposure category (based on two deaths). The white workers exhibited an erratic exposure-response relationship, with elevated risks in the 1st, 2nd and 5th categories, but a declining trend across the 2nd, 3rd and 4th categories.

Initially several specifications of exposure-response were examined within the family of log-linear models: 1a–1d above. Models with linear, square root, quadratic and log-transformed representations of cumulative chromium exposure performed about the same (Table 2, models 1–3, 4.1) but a model in which smoking cumulative exposure was log-transformed performed considerably better (Table 2, model 4.2). Use of a piece-wise linear specification for smoking exposure resulted in further improvement (Table 2, model 4.3). For a saturated model containing both the log-transformed and piece-wise linear smoking terms, the model deviance was almost identical to that with the piece-wise terms alone (1931.39 vs. 1931.57). Cubic splines applied to cumulative chromium exposure in log-linear models did not detect significant smooth departures from any of the specified models.

The linear model within the class of additive relative rate models (form 1e, above), with both chromium and smoking in the linear term without log transformation, suggested a superior fit (Table 3, model 1) compared to the best log-linear model with log-transformed exposures (Table 2, model 4.2). This was particularly evident in the contribution of the cumulative chromium exposure term ($\Delta[-2 \ln L]$= 15.5 vs. 13.8). As in the log-linear case, the fit of this model further improved when the piece-wise linear estimation was applied to cumulative smoking (Table 3, models 1 vs. 2; these are nested models: the sum of the smoking piece-wise terms equals smoking cumulative exposure). The contribution of the cumulative chromium exposure term also increased ($\Delta[-2 \ln L]$= 16.3). The negative intercept in these models results from comparing observed rates of lung cancer death against the national reference rates incorporated into the model. With smoking included in the model a negative intercept parameter describes the lowered mortality rates among nonsmoking cohort members as compared to the national population which includes smokers. Adding a product term for chromium and smoking cumulative exposures identified a negative but nonsignificant interaction in this model ($\Delta[-2 \ln L]$= 1.7, 2df). Allowing for some nonlinearity for the chromium exposure-response in this model (form 1f) did not significantly improve the fit (data not shown).

With only chromium exposure in the linear term and smoking exposure in the log-linear term of the linear relative rate model, the fit of the model was slightly reduced (Table 3, models 3, 4), but this specification permitted the smoking effect to be incorporated into the estimated background rate. Using this "final model" (Table 3, model 4) allowed a calculation of excess lifetime risk of hexavalent chromium exposure that did not distinguish smoking status; the estimate for each exposure level applied to all workers regardless of smoking. Furthermore, to calculate excess lifetime risk in the U.S. population using the model with a non-smoking baseline (Table 3, model 2) would require general population mortality data specific to age, race and smoking history, which are not available. With this model (Table 3, model 4), there was again a non-significant negative interaction between smoking and chromium cumulative exposures (interaction included in linear term); $\Delta[-2 \ln L]$= 2.1, 2df). On fit, this final model was essentially identical to the log-linear model using log-transformed chromium cumulative exposure (power model) and the piece-wise linear smoking terms (Table 2, model 4.3); the model deviances were 1931.60 and 1931.57 respectively. For the saturated model including both representations of chromium exposure together with the piece-wise linear smoking terms, the model deviance was 1931.26 (data not shown).

In the final model, the rate ratio estimated for 1 mg/m^3-yr cumulative exposure to hexavalent chromium was 2.44 (= estimated coefficient +1.0) with a 95% confidence interval of 1.54–3.83 ($\Delta[-2 \ln L]$= 15.1) (Table 4, model 1). At the mean cumulative exposure experienced by the lung cancer cases (0.28 mg/m^3-yr), the rate ratio estimate was 1.40; extrapolating to the maximum cumulative exposure of the cases (4.09 mg/m^3-yr), it was 6.9. The estimated rate ratio for 45 years of exposure at the OSHA PEL (0.10 mg/m^3) was 7.5.

The different exposure-response relationships by race observed in the categorical analysis (Table 1) were evident in regression models as strong chromium-race interactions. For the preferred linear relative rate model (Table 3, model 4; Table 4, model 1), addition of the race-chromium interaction term resulted in reductions in deviance of greater than 10.0, a highly statistically significant result

(X^2=10.6, p=0.001) (Table 4, model 1 vs. 2). This interaction was observed whether age, race and calendar time were adjusted by stratification (internal adjustment) or by using external population rates. The chromium exposure-response for white men was diminished with the interaction (RR=1.18, 95% CI=0.43–1.92, for 1 mg/m^3-yr cumulative exposure) but an overall lung cancer excess remained for that group.

Cumulative smoking was used in the final models despite absence of detailed smoking histories because, in comparison with models using a simple categorical classification (smoking at hire: yes, no, unknown), the models using cumulative smoking fit better. The changes in –2ln(Likelihood) for the cumulative smoking piece-wise linear terms versus the categorical terms (both with two degrees of freedom) were 28.42 and 25.83 respectively in the final model.

A significant departure of the estimated background rate (for unexposed workers) from the U.S. reference rate was observed with age and with race, but not with calendar time. The age effect corresponded to a reduction of 8–10% (RR=0.92, 0.90) for each 5 yr. of age (Table 4, models 1, 2 respectively). When chromium cumulative exposure was partitioned into three distinct latency intervals (5–9.99, 10–19.99, 20 or more yrs), there was no improvement in fit, and the chromium exposure interaction with race remained.

Estimates of Excess Lifetime Risk

Estimates of excess lifetime mortality from lung cancer resulting from up to 45 years of chromium exposure at concentrations 0.001–0.10 mg/m^3 were calculated based on the preferred model without the chromium-race interaction (Table 4, model 1). At 0.10 mg/m^3 (the current OSHA standard for total hexavalent chromium as CrO_3), 45 years of exposure corresponds to a cumulative exposure of 4.5 mg/m^3-yr and a predicted lifetime excess risk for lung cancer mortality of 255 per thousand workers (95% CI: 109-416) (Table 5). At 0.01 mg/m^3, one tenth of the current standard, 45 years of exposure corresponds to a lifetime excess risk of 31 per thousand workers (95% CI: 12–59). When the alternate log-linear model (Table 2, model 4.3) was used, the estimates of lifetime excess risk for lung cancer mortality were very similar (Table 5).

DISCUSSION

Model Choice

After consideration of a variety of log-linear and additive relative rate forms for modeling both smoking and chromium effects, a linear relative rate model with highly statistically significant exposure effects was selected to describe the lung cancer-chromium exposure response and for calculating excess lifetime risk. An equivalently fitting power model produced a slightly larger but less precise estimate of lifetime risk at the current PEL. The current findings are consistent with but not directly comparable to the results of Gibb et al. in the same population.[1] Gibb et al. used a log transformation of cumulative exposure: ln(cumX+d), where d was the smallest measured background exposure in the study (d=1.4×10^{-6} mg/m^3-yr). That metric was then used in a log-linear Cox regression model to estimate exposure-response.

Exposure Assessment

Extensive and systematic historical environmental air-sampling data were available covering the entire period of employment for this study population. Over 70,000 measurements collected between 1950 and 1985 were available for the exposure assessment.[1] These measurements were taken with the objective of characterizing typical rather than highest exposure scenarios, the latter being more commonly measured in industrial settings. The extent and quality of the exposure data are unique among chromium exposed populations and far exceed what is typically available for historical cohort studies of occupational groups. Although the exposure information in this study has clear advantages over previous studies of chromium workers, it also has its limitations including a lack of information on particle size and also on variability of exposures of individual workers having the same job title. Nonetheless, the measurement methodology used in the exposure surveys was generally consistent with what has been required by OSHA.

Potential Confounding

Although asbestos was identified as a potential exposure in this cohort, we do not believe that it is likely to have been an important confounder in this investigation. As in most plants of this period, asbestos was widely used and might have resulted in exposures among part of the workforce, particularly among skilled trades and maintenance workers (e.g., pipefitters, steamfitters, furnace or kiln repair and laborers). Asbestos exposure would not be expected to be generally correlated with chromium exposure in this population and thus should not have biased the internal exposure-response analysis. Furthermore, no cases of mesothelioma were reported.

Although cigarette smoking was controlled for in this analysis, there is a possibility of residual confounding by smoking because of the crudeness of the smoking data which pertained only to the time of hire. However, this was considerably more smoking information than is usually found in occupational studies. Furthermore, for smoking to have been a strong confounder in this analysis, its intensity (packs per day) would have needed to vary by level of hexavalent chromium exposure concentration. Our assumption that smokers started at 18 years of age and continued until the end of follow-up permitted the estimation of a time-dependent smoking cumulative exposure. A piece-wise linear fitting of smoking cumulative exposure considerably improved model fit. It also indicated that cumulative smoking greater than 30 pack-years, where smoking misclassification would be greater due to unknown lifetime smoking history, was not a significant predictor of further increased risk (Tables 2–4). Observing a plateau in the smoking cumulative exposure response was consistent with the pattern observed in previous studies of smoking.[16] When smoking was modeled using the piece-wise linear terms, the parameter estimate of the chromium affect was increased by 10 percent, compared with use of the categorical specification for smoking.

Exposure-Race Interaction

As indicated above, an exposure-race interaction was observed in our analyses. Because the source of this interaction is unknown, we chose not to include an interaction term in the final risk assessment model. Discussion in the scientific literature of differential cancer rates and risks by race, focuses almost exclusively on socioeconomic, occupational and life-style risk factor differences and diagnostic bias.[17] We are not aware of examples establishing that occupational lung cancer susceptibility varies between African-Americans (majority of the nonwhite study population) and white Americans, and thus, in our opinion, it is unlikely that the chromium-race interaction that we observed has a biological basis. More plausible explanations include, but are not limited to: misclassification of smoking status, misclassification of chromium exposures, or chance. We have no evidence to support any of these explanations, however, and we believe that the exposure-response relationship derived for the entire study population provides the best basis for predicting risk.

Excess Lifetime Risk and Prior Risk Assessments

Our analysis predicts, based on the preferred model (Table 4, Model 1), that workers exposed at the current OSHA PEL of 0.1 mg/m³ for 45 years will have approximately 25% excess risk of dying from lung cancer due to their exposure. Very few workers in this study had cumulative exposures corresponding to 45 years at the PEL, and thus our estimates of risk are based on an upward extrapolation from most of the data (less than 2% of person-years but 10% of lung cancer deaths occurred with cumulative exposures greater than 1.0 mg/m³-yr (Table 1)). However, even workers exposed at one tenth of the PEL (i.e., to 0.01 mg/m³) would experience 3% excess deaths (Table 5).

There have been several other risk assessments for hexavalent chromium exposure and lung cancer, and it is informative to compare the predictions from these assessments with those from this investigation. In a 1995 review that included an earlier assessment of this cohort and the Mancuso study[2], OSHA identified point estimates of lifetime risk at the current PEL (0.1 mg/m³) in the range of 88 to 342 per thousand.[18] Conversion[¶] of a U.S. Environmental Protection Agency risk assessment[19] for ambient chromium exposures based on the Mancuso study, to predict occupational risks, produces estimated lifetime risks of 90 per thousand for 45 years of exposure to 0.1 mg/m³. The State of California EPA published a risk assessment based on best estimates also from the Mancuso study with different assumptions about chromium exposures which, when converted for occupational exposures, result in a predicted lifetime occupational risk of 90 to 591 per thousand for exposure at the PEL.[20] Thus the predicted occupational risks for lifetime exposure to the current OSHA PEL developed from these previous risk assessments, which ranged from 88 to 591 per 1000 workers, are quite consistent and bracket the estimate presented in this paper of 255 per 1000 workers. Thus the estimates of risk for 45 years of exposure at the current OSHA PEL from previous risk assessments are all within a factor of 3 of the estimates provided in this paper, which is reasonably consistent given the uncertainties involved in the risk assessment process.

CONCLUSION

This study clearly identifies a linear trend of increasing risk of lung cancer mortality with increasing cumulative exposure to water-soluble hexavalent chromium. Our analysis predicts that exposure

---

[¶]The U.S. and California EPA risk assessments reported unit lifetime risks for hexavalent chromium exposure of 0.012 (per µg/m³). To convert this to occupational risk and exposure to chromium as chromic acid, the unit risk was multiplied by 240/365 days to correct for differences in number of days exposed per week, 8/24 to correct for number of hours per day exposed, 45/70 to correct for number of years of exposure, and by 0.52 to convert from chromium to CrO_3.

at the current OSHA PEL for hexavalent chromium permits a lifetime excess risk of lung cancer death that exceeds 1 in 10. Exposures at $1/10^{th}$ the PEL (0.01 mg/m^3 or 10 μg/m^3, as CrO_3), would confer a 3 per hundred lifetime risk. The risk estimates from this analysis are consistent with those of other assessments.

ACKNOWLEDGEMENTS

The authors would like to acknowledge the contributions of the following in providing helpful commentary: Drs. John Bailer, Kyle Steenland, Christopher Portier, Chao Chen, David Dankovic and Paul Pinsky, Ms. Caroline Freeman, and two journal referees.

REFERENCES

1. Gibb HJ, Lees PSJ, Pinsky PF, Rooney BC. Lung cancer among workers in chromium chemical production, Am J Ind Med 2000; *38*:115–126.
2. Mancuso TF. Chromium as an industrial carcinogen: Part I. Am J Ind Med 1997; *31*:129–139.
3. Luippold RS, Mundt KA, Panko J, Liebig E, Crump C, Crump K, Paustenbach D, Proctor D. Lung cancer mortality among chromate production workers. Occup Environ Med 2003; *60*:451–457.
4. U.S. Occupational Safety and Health Administration. 2000. 29CFR1910.1000, Table Z-2
5. American Conference of Government and Industrial Hygienists. 1999. 1999 TLVs and BEIs. Cincinnati, OH: ACGIH.
6. NIOSH. 1975 Criteria for a recommended standard: Occupational exposure to chromium(VI). U.S. Department of Health and Human Services, Public Health Service, Centers for Disease Control, National Institute for Occupational Safety and Health, Washington D.C. HEW (NIOSH) publication no. 76–129.
7. Hayes RB, Lilienfeld AM, Snell LM. Mortality in chromium chemical production workers: a prospective study. Int J Epidemiol 1979; *8*:365–374.
8. Park RM, Bailer A.J, Stayner LT, Halperin W, Gilbert SJ. An alternate characterization of hazard in occupational epidemiology: years of life lost per years worked. Am J Ind Med 2002; *42*:1–10.
9. Monson R. USDR97 rates. Department of Environmental Health, Harvard School of Public Health, Boston, MA, 1997.
10. Preston DL, Lubin JH, Pierce DA, et al. Epicure Users Guide. Seattle, WA: Hirosoft International Corp., 1993.
11. Chambers JM, Hastie TJ. Statistical Models in S. Pacific Grove, CA: Wadsworth, Inc., 1992.
12. Mathsoft, Inc. S-Plus 4 Guide to Statistics. Seattle, WA: Mathsoft, Inc., 1997.
13. Biological Effects of Ionizing Radiation (BEIR) IV. Health risks of radon and other internally deposited alpha-emitters. Committee on the Biological Effects of Ionizing Radiation, Board of Radiation Effects Research, Commission on Life Sciences, National Research Council. National Academy Press, Washington, D.C., 1988.
14. National Center for Health Statistics. Vital statistics of the United States. Volume II—Mortality Part A. Washington: Public Health Service, 1996. DHHS Publication No. (PHS) 96–1101.
15. Braver ER, Infante P, Chu K. An analysis of lung cancer risk from exposure to hexavalent chromium. Teratogenesis, Carcinogenesis, and Mutagenesis 1985; *5*:365–378.

16. Vineis P, Kogevinas M, Siminato L, Brennan P, Boffetta P. Leveling off of the risk of lung and bladder cancer in heavy smokers: an analysis based on multicentric case-control studies and a metabolic interpretation. Mutat Res 2000; *463*:103–110.
17. Schottenfeld D, Fraumeni JF, eds, Cancer Epidemiology and Prevention, 2nd Edition, Oxford University Press, New York, 1996.
18. OSHA. Ex.13-5, Docket # H-054a - Room N2526, U.S. Dept of Labor, OSHA, 200 Constitution Ave., NW, Washington DC, 20210; May, 1995.
19. EPA. Health assessment document for chromium. EPA-600/8-83-14F, U.S. Environmental Protection Agency, Washington, D.C.
20. Cal EPA: http://www.oehha.ca.gov/air/toxic_contaminants/pdf1/ hexavalent%20chromium.pdf

Table 1. SMRs for lung cancer in five strata of cumulative exposure to hexavalent chromium (mg/m³-yr, as CrO^3)

	Cum. exposure	P-yrs	Obs	Exp	SMR	95%CI	SRR	95% CI
All workers (includes 4 lung cancer cases with unknown race)								
1	[0.0000-0.0282]	51348	72	47.93	1.50	1.18-1.88	1.0	—
2	[0.0282-0.0944]	7837	14	7.64	1.83	1.03-2.97	1.29	0.69-2.22
3	[0.0944-0.3715]	6859	12	6.09	1.97	1.06-3.31	1.38	0.71-2.46
4	[0.3715-1.0949]	3841	12	5.13	2.34	1.25-3.93	1.70	0.87-3.03
5	[1.0949-5.2600]	950	12	1.90	6.32	3.39-10.60	4.53	2.32-8.13
	Total		122	68.68	1.78	1.50-2.11		
White workers								
1	[0.0000-0.0282]	27962	45	27.32	1.65	1.21-2.18	1.0	—
2	[0.0282-0.0944]	4088	12	3.93	3.06	1.64-5.13	1.85	0.92-3.43
3	[0.0944-0.3715]	3409	6	2.98	2.01	0.80-4.08	1.21	0.46-2.70
4	[0.3715-1.0949]	2188	4	3.22	1.24	0.39-2.89	0.86	0.26-2.15
5	[1.0949-5.2600]	495	4	1.03	3.87	1.20-8.98	2.55	0.76-6.38
	Total		71	38.48	1.85	1.45-2.31		
Nonwhite workers								
1	[0.0000-0.0282]	16384	24	16.16	1.49	0.97-2.16	1.0	—
2	[0.0282-0.0944]	3118	2	3.31	0.60	0.10-1.86	0.41	0.07-1.41
3	[0.0944-0.3715]	3125	5	2.93	1.71	0.61-3.67	1.24	0.41-3.04
4	[0.3715-1.0949]	1589	8	1.87	4.23	1.96-7.98	3.03	1.24-6.64
5	[1.0949-5.2600]	434	8	0.85	9.41	4.30-17.5	6.76	2.76-15.0
	Total		47	25.12	1.87	1.39-2.46		
Workers (unknown race)								
1	[0.0000-0.0282]	7002	3	4.45	0.67	0.17-1.75	1.0	—
2	[0.0282-0.0944]	631	0	0.40	0.00	0.00-4.78	0.00	0.00-55.1
3	[0.0944-0.3715]	326	1	0.17	5.73	0.33-25.2	11.2	0.52-117.7
4	[0.3715-1.0949]	64	0	0.04	0.00	0.00-47.8	0.00	—
5	[1.0949-5.2600]	22	0	0.01	0.00	0.00-143.9	0.00	—
	Total		4	5.08	0.79	0.24-1.83		

Appendix A

Table 2. Log-linear Poisson regression model forms with different specifications for chromium exposure

	Models[1]	Exposure effect ln(RR)	Deviance	Δ-2ln(L) (df)[2]	p
1	Log-linear (1a)				
	Intercept	0.0980	1946.79		
	Cumulative smoking	0.0225			
	Cumulative chromium-6	0.4950		9.813 (1)	0.0017
2	Log-square root (1b)				
	Intercept	0.0073	1946.49		
	Cumulative smoking	0.0211			
	Sqrt(Cum. chromium-6)	0.7884		10.12 (1)	0.0015
3	Log-quadratic (1c)				
	Intercept	0.0942	1945.10		
	Cumulative smoking	0.0207			
	Cumulative chromium-6	0.8935		11.51 (2)	0.0032
	Square(Cum. chromium-6)	-0.1280			
4	Power (1d)				
	4.1 Intercept	0.0850	1945.74		
	Cumulative smoking	0.0211			
	Ln(Cum. chromium-6+1)	1.048		10.87 (1)	0.0010
	4.2 Intercept	-0.9661	1934.67		
	Ln(Cumulative smoking+1)	0.5139			
	Ln(Cum. chromium-6+1)	1.178		13.81 (1)	0.0002
	4.3 Intercept	-0.7804	1931.57		
	Smoking < 30 pack-years[3]	0.0613			
	Smoking > 30 pack-years[4]	-0.0053			
	Ln(Cum. chromium-6+1)	1.253		15.08 (1)	0.0001

[1]Models include race as a categorical (3-level) variable, and allow for a linear departure of the baseline from reference rates with age, centered at 50 years; cumulative smoking is in packs/day-years (lag=5 years), and cumulative chromium as CrO_3 is in mg/m^3-years (lag=5 years)
[2]Change in -2ln(L) for terms involving chromium, basis for chi-sq statistical test
[3]Specified as: minimum(cumulative smoking, 30 pack-yrs)
[4]Specified as: maximum(cumulative smoking—30 pack-yrs, 0)

Table 3. Linear relative rate Poisson regression model forms for chromium exposure

	Models[1]	Estimate of exposure effect	Deviance	Δ-2ln(L) (df)[2]	p
Model (1e): linear relative rate in smoking and chromium; loglinear rate in race, age					
1	Intercept	-1.377	1933.71		
	Cumulative smoking	0.2833			
	Cumulative chromium-6	8.994		15.46 (1)	< 0.0001
2	Intercept	-1.636	1930.51		
	Smoking < 30 pack-years[3]	0.402			
	Smoking > 30 pack-years[4]	0.031			
	Cumulative chromium-6	10.95		16.25 (1)	< 0.0001
Model (1e): linear relative rate in chromium; log-linear rate in smoking, race, age					
3	Intercept	-0.965	1934.75		
	Ln(Cumulative smoking+1)	0.513			
	Cumulative chromium-6	1.269		13.73 (1)	0.0002
4	Intercept	-0.786	1931.60		
	Smoking < 30 pack-years	0.061			
	Smoking > 30 pack-years	-0.006			
	Cumulative chromium-6	1.444		15.05 (1)	0.0001

[1]Models include race as a categorical (3-level) variable, and allow for a linear departure of the baseline from reference rates with age, centered at 50; cumulative smoking is in packs/day-years (lag=5 years), and cumulative chromium as CrO_3 is in mg/m^3-years (lag=5 years)
[2]Change in -2ln(L) for terms involving chromium, basis for chi-sq statistical test
[3]Specified as: minimum(cumulative smoking, 30 pack-yrs)
[4]Specified as: maximum(cumulative smoking—30 pack-yrs, 0)

Table 4. Final Poisson regression model and model including chromium-race interaction (lag=5 years)

	Model[1]	Estimate of exposure effect	Δ-2ln(L) (df)[2]	SMR[3]	RR[4]	RR, 95% CI	Reference[5]
Model (1e): linear relative rate in chromium, log-linear in smoking, race, age							
1	Race: nonwhite men	-0.786		0.46			US NW men
	Race: white men	-0.901		0.41			US W men
	Race: unknown 5	-1.515		0.22			US men
	Age	-0.082			0.92		5 yr of age
	Smoking < 30 pack-years[6]	0.061			1.06	1.04,1.09	pack-yrs
	Smoking > 30 pack-years[7]	-0.006			0.99	0.97,1.01	pack-yrs
	Cumulative chromium-6	1.444	15.05 (1)		2.44	1.54,3.83	mg/m^3-yr
(Model deviance: -2ln(L) =1931.60)							
2	Race: nonwhite men	-1.121		0.33			US NW men
	Race: white men	-0.834		0.43			US W men
	Race: unknown	-1.557		0.21			US men
	Age	-0.100			0.90		5 yr of age
	Smoking < 30 pack-years	0.064			1.07	1.04,1.07	pack-yrs
	Smoking > 30 pack-years	0.001			1.00	0.98,1.02	pack-yrs
	Cum. chrom-6, NW men	4.312	25.63 (2)		5.31	2.78,10.1	mg/m^3-yr
	Cum. chromium-6, W men	0.176			1.18	0.43,1.92	mg/m^3-yr
(Model deviance: -2ln(L) =1921.02)							

[1] The baseline from reference rates with age, centered at 50 years; cumulative smoking is in packs/day-years (lag=5 years), and cumulative chromium as CrO$_3$ is in mg/m^3-yr (lag=5 years)
[2] Change in -2ln(L) for terms involving chromium, basis for chi-sq statistical test
[3] Estimated SMR adjusted for terms in model, i.e., unexposed, in race strata, at age 50
[4] R, relative rate, for cum. exposure of 1.0 mg/m^3-yr
[5] W=white, NW=nonwhite; for unknown race, US rates applied in proportion to those with known race by year of hire
[6] Specified as: minimum(cumulative smoking, 30 pack-yrs)
[7] Specified as: maximum(cumulative smoking—30 pack-yrs, 0)

Table 5. Excess lifetime risk of lung cancer mortality for specified concentrations of hexavalent chromium (as CrO_3) assuming 45 year exposure

Hexavalent chromium exposure (as CrO_3, mg/m³)	Excess lifetime risk			
	Linear relative rate model[1]		Log-linear model[2]	
	Excess risk[3]	95% CI	Excess risk	95% CI
0.000	0.000	—	0.000	—
0.001	0.003	0.001–0.006	0.003	0.001–0.004
0.002	0.006	0.003–0.012	0.005	0.003–0.008
0.005	0.016	0.006–0.030	0.014	0.007–0.020
0.010	0.031	0.012–0.059	0.028	0.013–0.043
0.020	0.060	0.023–0.113	0.057	0.025–0.093
0.050	0.141	0.057–0.251	0.145	0.056–0.264
0.100[4]	0.255	0.109–0.416	0.281	0.096–0.516

[1]Based on Table 4, model 1
[2]Based on Table 2, model 4.3
[3]Probability of chromium-attributable lung cancer death in lifetime with exposure starting at age 20 and lasting up to 45 years (calculated through age 85)
[4]OSHA PEL for total hexavalent chromium

www.ingramcontent.com/pod-product-compliance
Lightning Source LLC
Chambersburg PA
CBHW080249180526
45167CB00006B/2468